Advances in Soil Science

Advances in Soil Science

B. A. Stewart, Editor

Advances in Soil Science

Volume 5

Edited by B.A. Stewart

With Contributions by
S.K. De Datta, M.P.W. Farina, R. Lal, P.K. Sharma,
D.E. Smika, M.E. Sumner, P.W. Unger, and G. Zitong

With 74 Illustrations

Springer-Verlag
New York Berlin Heidelberg London Paris Tokyo

B.A. Stewart
USDA Conservation & Production Research Laboratory
Bushland, Texas 79012
U.S.A.

The Library of Congress has cataloged this
serial publication as follows:
Advances in soil science.—Vol. 1—New York: Springer-Verlag,
c1985-
v.:ill.; 25 cm.
ISSN 0176-9340 = Advances in soil science.

1. Soil science—Collected works.
S590.A38 631.4'05—dc19 85-644892
AACR 2 MARC-S
Library of Congress [8512]

Softcover reprint of the hardcover 1st edition 1986

Typeset by Ampersand Publisher Services, Inc., Rutland, Vermont.

9 8 7 6 5 4 3 2 1

ISBN- 3:978-1-4613-8662-9 e-ISBN-13:978-1-4613-8660-5
DOI: 10.1007/978-1-4613-8660-5

Preface

The world needs for food and fiber continue to increase. Population growth in the developing countries peaked at 2.4% a year in 1965 and has fallen to about 2.1%. However, in many developing countries almost half the people are under 15 years of age, poised to enter their productive and reproductive years. The challenges to produce enough food for this growing population will remain great. Even more challenging is growing the food in the areas of greatest need. Presently the world has great surpluses of food and fiber in some areas while there are devastating deficiencies in other areas. Economic conditions and the lack of suitable infrastructure for distribution all too often limit the alleviation of hunger even when there are adequate supplies, sometimes even within the country itself.

World hunger can be solved in the long run only by increasing crop production in the areas where the population is growing most rapidly. This will require increased efforts of both the developed and developing countries. Much of the technology that is so successful for crop production in the developed countries cannot be utilized directly in the developing countries. Many of the principles, however, can and must be adapted to the conditions, both physical and economic, of the developing countries.

This series, *Advances in Soil Science,* was established to provide a forum for leading scientists to analyze and summarize the available scientific information on a subject, assessing its importance, and identifying additional research needs. Most importantly, the contributors were asked to develop and identify principles that have practical applications to both developing and developed agricultures. It is not the purpose of the series to report new research findings because there are many excellent scientific journals for that need. Communications in scientific journals, however, are generally restricted to short and technical presentations. Therefore, *Advances in Soil Science* fills a gap between the scientific journals and the comprehensive reference books, in which scientists can delve in depth on a particular subject relating to soil science. The positive response to the first volumes of the series shows that such a gap did indeed exist.

The ultimate aim of the series is to stimulate action: action to determine where there are arable soils, action to develop technology for more efficient crop production on these soils, action to reduce the risk of degrading these soil resources, and action to determine on which soils our research efforts should be concentrated. Without such action, the task of producing adequate food in the future may simply be too great.

There are many audiences to reach. While intended primarily for scientists and students of soil science, this series also provides technical information for anyone interested in our natural resources and man's influence on these resources. The reviews are written by leading scientists from many countries and therefore provide the reader with information from a wide array of conditions. Such information will be particularly useful to professionals working in areas with developing agricultures because the reviews summarize and assess the significance of the technical literature.

B.A. Stewart

Contents

Contributors

SURAJIT K. DE DATTA, International Rice Research Institute, Manila, Philippines
MART P.W. FARINA, The Grain Crops Research Institute, Department of Agriculture, Pietermaritzburg, South Africa
R. LAL, International Institute of Tropical Agriculture, Ibadan, Nigeria
PRADEEP K. SHARMA, International Rice Research Institute, Manila, Philippines
DARRYL E. SMIKA, USDA Agricultural Research Service, Akron, Colorado 80720, U.S.A.
MALCOLM E. SUMNER, University of Georgia, Department of Agronomy, Athens, Georgia 30602, U.S.A.
PAUL W. UNGER, USDA Agricultural Research Service, Bushland, Texas 79012, U.S.A.
GONG ZITONG, The Institute of Soil Science, Academic Sinica, Nanking, People's Republic of China

Soil Surface Management in the Tropics for Intensive Land Use and High and Sustained Production

R. Lal*

---

*International Institute of Tropical Agriculture, Ibadan, Nigeria.

Advances in Soil Science, Volume 5

I. Introduction

The tropics cover about 40% of the earth's surface. With few exceptions, most soils of these regions are characterized by low yields and are susceptible to a rapid decline in soil productivity. The latter is associated with degradation of soil quality by intensive land use. Because of the severe soil constraints, national yield averages in tropical Africa rarely show an upward trend despite improved cultivars and agrochemical inputs. Agricultural scientists and planners are increasingly concerned about the severe lag in food production, in view of the ever-increasing demand. Consequently the potential and capability of soils to sustain high production with intensive management and continuous cultivation are being questioned. The important issues to be resolved are: Are soils of the tropics capable of sustaining high and economical production? If yes, with what methods of soil and crop management and by which farming systems? What is the approach to be adopted in transforming inefficient, low-production, and subsistence traditional agriculture into a sustainable commercial enterprise? Can this approach provide the basic necessities of life and meet the aspirations of modern society and yet maintain ecological stability and preserve the natural resource? The objective of this report is to describe soil resources of the tropics, assess their potential and constraints, and review state of the art knowledge concerning the improved systems of management for high production.

II. Land, Rainfall, and Vegetation

A. Arable Land

About 64% of the 3 billion hectares of potentially cultivable land areas lies in Asia, Africa, and South America and only 37% of the land in the tropics has ever been cultivated (FAO, 1981). Estimates of potentially cultivable area in the developing tropical countries alone range from 1145 million to 1648 million hectares. Norse (1979) estimated the total cultivable land area to be 502 and 1164 million hectares in the humid and dry tropics, respectively. Out of this the presently cultivated area is merely 86 million hectares in the humid tropics and 424 million hectares in the dry tropics. There are also some sparsely inhabited regions in the tropics

Table 1a. Land and water resources in the tropics

Region	*Presently cultivated* ($\times 10^6$ *ha*)	*Total potentially cultivable* ($\times 10^6$ *ha*)
Wet tropics	86	502
Dry tropics	424	1164
World total	1447	3139

Reprinted with permission from Norse, in: M.R. Biswas and A.K. Biswas (eds.), *Food, Climate and Man*, p. 20. Copyright 1979 by John Wiley & Sons, Inc.

that can be developed for food crop production (Table 1a). For example, potentially cultivable area in the Amazon is 450 million hectares. If properly developed and managed, these regions can support a high population. Potentially cultivable land areas also exist in Central and West Africa and in Southeast Asia (Table 1b).

With water being the limiting factor, genuine efforts have been made to increase the percentage of arable land that can be irrigated. The FAO (1984) estimated that in the decade ending in 1982, the irrigated land area increased from 7.4 to 8.7 million hectares in Africa, from 6.0 to 7.7 million hectares in tropical America, and from 114.1 to 134.8 million hectares in Asia (Table 2). In spite of this, however, the arable land area per agricultural worker has decreased in all regions of the tropics except Latin America. The arable land area per agricultural worker is below 1 ha in China and less than 1.5 ha in Africa (Table 3). The strategy for improving farm production must therefore address itself to the problems of small landholders with low financial resources.

B. Rainfall

The mean annual rainfall in tropical regions varies widely, from as low as 250 mm or less to more than 6000 mm. In some regions of West Africa and northeast India, annual rainfall sometimes exceeds even 10,000 or

Table 1b. Forest area in the tropics that can be developed for agriculture

Region	*Closed forest* ($\times 10^6$ *ha*)	*Open woodlands* ($\times 10^6$ *ha*)
Latin America	680	33
Asia	410	15
Africa	190	6
Oceania	89	10
World total	2657	20

Adapted from Barney, *Global 2000 Report to the President of the USA*. Copyright 1980, Pergamon Press.

Table 2. Irrigated and total land area

Region	1972 Irrigated (× 10^6 ha)	1972 Total (× 10^6 ha)	1982 Irrigated (× 10^6 ha)	1982 Total (× 10^6 ha)
Africa	7.4	171.8	8.7	184.0
South America	6.0	115.3	7.7	139.4
Asia	114.1	447.6	134.8	456.8
Europe	23.7	373.5	33.6	373.0
North and Central America	21.8	266.7	22.3	273.3

Reprinted with permission by the Food and Agriculture Organization of the United Nations, 1984, from *Agriculture: Toward 2000.*

12,000 mm. As a general rule the equatorial belt (the region about 5° north and south of the equator) receives the maximum rainfall. In West Africa, Kowal and Kassam (1978) observed a regular north–south gradient in the rainfall pattern. In this region, the rainfall decreases with increasing distance away from the equator:

$$P\ (\text{mm}) = 2470 - 130.9LA - 0.6LO;\ r = 0.91$$

where P is the mean annual rainfall, and LA and LO refer to the latitudinal and longitudinal coordinates. Kowal and Kassam observed that an average decrease in rainfall is 131 mm per degree latitude northward or 1.18 mm km^{-1} inland from the ocean. Similar patterns are observed for tropical Australia (Fitzpatrick, 1965).

Although the total annual rainfall received can be adequate, the problems to agriculture are posed by its variability, distribution, and

Table 3. Arable land per agricultural worker

Region	Hectares 1967–1970	Hectares 1982
Africa	1.61	1.48
Far East	1.00	0.94
Near East	2.67	2.40
Latin America	3.99	4.45
China	0.40	0.38
East Europe and USSR	5.35	7.66
North America	54.51	92.33
Western Europe	4.15	6.12
Oceania	74.49	103.98

Reprinted with permission by the Food and Agriculture Organization of the United Nations, 1984, from *Agriculture: Toward 2000.*

irregularities. Tropical rains are supposed to have high spatial and temporal variability and therefore vary markedly from place to place and year to year. Riehl (1979) observed that the highest and the lowest rainfall received at Bombay, India, during 110 years of recordings was double and half the annual means, respectively. For Honolulu, Hawaii, with a mean rainfall of 71 cm, 85 years of records indicated a maximum rainfall of 142 cm and a minimum of 25 cm. The annual rainfall received from 1947 to 1983 at Maroua, Cameroon, shown in Figure 1, indicates drastic deviations from the long-term average of 784 mm $year.^{-1}$

Tropical rains are often localized and spotty, and the variability among locations within short distances from one another can be high. Among the factors responsible for this high spatial variability are topography and relief, slope aspect, and directional storms. Nwa (1977) reported variability of 10 to 15% among four gages distributed over a watershed of about 40 ha. In Bihar, India, Sreenivasan (1971) computed spatial variability in monthly rainfall for the 32 stations and observed that the variations in space were of a much higher order than the variations in time.

In general, tropical rains are less effective than temperate rains. Because of high intensity, a relatively high proportion is lost as runoff. High intensities of 150 to 200 mm hr^{-1} have been observed for a relatively short duration of 5 to 10 minutes (Hudson, 1976; Lal, 1976). the losses from evaporation are also high because of the prevalence of high temperatures. Because of favorable day/night temperatures plant growth rates are generally high, leading to high consumptive rates. Consequently, seasonal crops suffer from drought within a few days after a heavy rain. If the probability of a 5- to 7-day rainless period is high, crop yield is adversely affected.

Growing Season

On the basis of rainfall distribution, four moisture regimens are commonly observed in the tropics (Le Houerou and Povov, 1981):

1. No rainy season/permanent drought
2. A single rainy season/one dry season
3. Two rainy seasons/two dry seasons
4. No dry season/permanent rains

Regions with monomodal rainfall distribution are called "tropical" and those with bimodal pattern are "equatorial" (De Martonne, 1926). In the tropics, the springs or the beginnings of the monsoons are hot and dry, in contrast to the cold and wet springs in the temperate region. In regions with prolonged dry seasons, low soil moisture and high soil temperature are the critical factors that determine the success or failure of early planted crops. It is important, therefore, to clearly define the beginning of the growing season, which apparently coincides with the onset of "assured rains" or the "wet month." Various definitions exist of what constitues a "wet month." Some researchers consider percentage variability in rainfall as a diagnostic criterion (Kenworthy and Glover, 1958). Dry months generally have a variability exceeding 70%. For the semi arid tropics, Virmani (1975) defined start of the growing season as an event with 20 mm of rain falling on one or two successive days. Benoit (1977) defined the start of the growing season as the date when accumulated rainfall exceeds and remains greater than one-half of the potential evapotranspiration for the remainder of the growing season provided that no dry spell longer than 5 days occurs immediately after this date. It has been observed that in West Africa mean start of the growing season delays with increase in distance from the equator (Kowal and Kassam, 1978).

Start of rains (decade) $= -1.16 + 1.34\,LA + 0.70\,LO;\ r = -0.94$

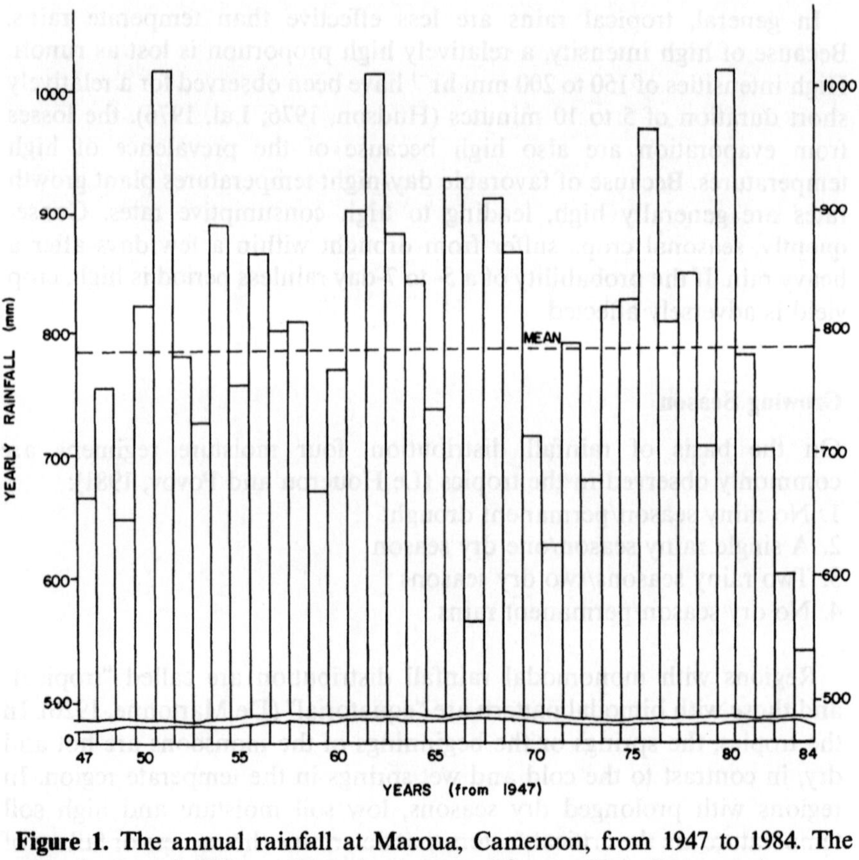

Figure 1. The annual rainfall at Maroua, Cameroon, from 1947 to 1984. The climatic mean is 784 mm. (Data by courtesy Director, IRA, Cameroon.)

Figure 2. Distribution of tropical rainforest in the world (N. Myers, 1980). Reprinted with permission by the National Science Foundation.

where *LA* and *Lo* refer to the latitude and longitude. Similar trends exist in the onset of monsoons in the Indian subcontinent. In Australia, Fitzpatrick *et al.* (1967) estimated the length of the growing season on the basis of rainfall and soil-moisture availability.

D. Vegetation

Two agriculturally important vegetation types are tropical forest and tropical savanna.

Tropical forests occupy the equatorial green belt (Figure 2). In this ecology the rainfall often exceeds the evapotranspiration for at least 8 months in a year, with a mean annual rainfall of 1400 mm and above. This vegetation zone is characterized by high species diversity. It is not

Plate 1. A tropical rainforest at Okomu, Southern Nigeria.

Plate 2. *Imperata cylindrica* in the abandoned coffee land in Brazil.

uncommon to find 40 to 100 species growing in 1 ha. Some trees can reach a height of 50 to 60 m, with multistorey canopy (Plate 1). The total biomass of tropical rainforest amounts to 350 to 450 t ha^{-1}, with a leaf area index of 12 to 15; this gives a gross production of 120 to 150 t ha^{-1} annually, corresponding to a primary production of 30 to 35 t ha^{-1} of which 10 to 12 t ha^{-1} are accounted for by litter (Walter, 1979). The estimates of tropical forest reserves vary widely. According to UNESCO (1978) the total forested land in the tropics is about 1 billion hectares (Table 4a). In contrast, Postel (1984) estimated forest resources of the tropics to be 2969 million hectares, or 60.7% of the world's forest reserves (Table 4b).

Table 4a. Estimates of forest reserves in the tropics

Region	*Evergreen rainforest (× 10^6 ha)*	*Semidecidous forest (× 10^6 ha)*	*Total (× 10^6 ha)*
Southeast Asia	124.9	3.3	128.2
Tropical Africa	83.25	55.5	138.75
Tropical America	557.0	168.0	725.00
World total	765.15	226.8	1021.95

Modified version of Summary and Table 2 from TROPICAL FOREST ECOSYSTEM.

Table 4b. Forest resources of the tropics

Region	*Closed forest (× 10^6 ha)*	*Open woodland (× 10^6 ha)*	*Forest fallow and shrubland (× 10^6 ha)*	*Total Total (× 10^6 ha)*
Tropical America	679	217	317	1212
Tropical Africa	217	486	609	1312
Tropical Asia	305	31	109	445
Total	1201	734	1035	2969
Percentage of the world	44.9	63.3	98.1	

Adapted from Postel, 1984.

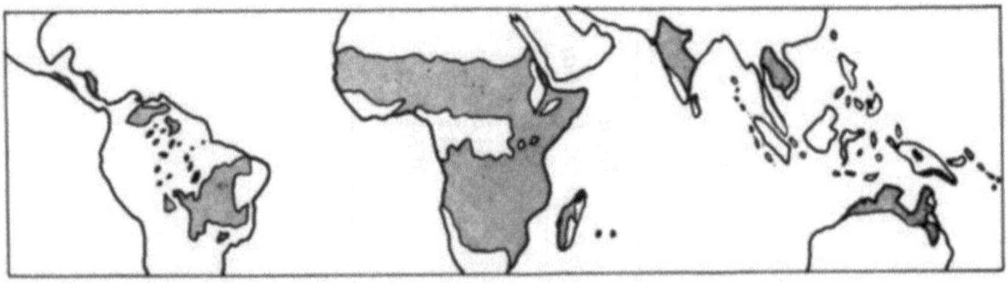

Figure 3. Distribution of tropical savannas, in: Golley F. B. (ed), *Tropical Savannas.* Copyright 1983. Elsevier Science Publishers.

The transitional zone between rainforest and tropical savanna is occupied by deciduous forests. The duration of the leafless state of the trees increases with the length of the dry season. Savanna vegetation occurs extensively in the semiarid regions of Africa, Asia, Australia, and South America (Figure 3). In areas of marginal rainfall, forest vegetation is gradually replaced by tree or grass savanna. These formations are often termed "derived savanna" or "manmade" savanna. Some forested regions on marginal soils are converted to grass savanna and often colonized by *Imperata cylindrica* (Plate 2). Vast areas of *Imperata*-infested lands occur in Sumatra, Indonesia, West Africa, and tropical America.

The savanna ecology is characterized by a mean annual rainfall of 600 to 1400 mm, the seasonality of rainfall pattern, with the less effective and highly variable rainfall during the growing season, and a prolonged dry season. There are two types of savanna vegetation: grass and woody. These two vegetation types are usually antagonistic. While the grasses have intensive root systems confined to the surface layer, the woody species have an extensive root system penetrating deeper into the subsoil. The net primary productivity of the savanna ecosystem is about 2.5 times less than that of the rainforest. The low net annual productivity of the savanna is attributed mostly to water deficit.

III. Soil Resources of the Tropics

Major soils of the tropics are Oxisols, Aridisols, Alfisols, Ultisols, Inceptisols, Entisols, and Vertisols (Table 5). Similar to soils of the temperate zone, soils of the tropics are highly diverse and variable depending on geology, geomorphology, vegetation, and the rainfall pattern. These soils can be broadly subdivided into three major categories: (i) soils containing predominantly low-activity clays, (ii) soils of mixed mineralogy and those containing high-activity clays, and (iii) soils of volcanic origin.

Table 5. Land area of the major soils of the tropics

	World land area		*Land area in the tropics*	
Soil order	*× 10^6 ha*	*%*	*× 10^6 ha*	*%*
Alfisols	1,730	13.1	800	16.2
Aridisols	2,480	18.8	900	18.4
Entisols	1,090	8.2	400	8.2
Histosols	120	0.9	—	—
Inceptisols	1,170	8.9	400	8.3
Mollisols	1,130	8.6	50	1.0
Oxisols	1,120	8.5	1,100	22.5
Spodosols	560	4.3	—	—
Ultisols	730	5.6	550	11.2
Vertisols	230	1.8	100	2.0
Mountains	2,810	21.3	600	12.2
Total	13,170	100.0	4,900	100.0

Reprinted with permission from Buringh, *Introduction to the Study of Soils,* p. 83. Copyright 1979 by the Centre for Agricultural Publishing and Documentation, Wageningen, The Netherlands.

Soils of the rainforest and semideciduous forest regions are those containing low-activity clays. These soils are characterized by clay fractions mostly composed of kaolinite and halloysite with hydrous oxides of iron and aluminum. Soil orders Alfisols, Ultisols, and Oxisols of this category comprise about 50% of the land surface of the tropics and about 27% of the world land area.

Soil orders containing high-activity clays occur mostly in the semiarid and arid regions and include Aridisols, Vertisols, Mollisols, Inceptisols, and Entisols. The predominant soils of the arid tropics—Aridisols and Vertisols—occupy about 20% of the tropical land area. Recently developed Inceptisols and Entisols are relatively fertile soils and occupy about 16.5% of the tropical land area. Soils of volcanic origin (Andosols) are the most productive and occur in Central America, Java, and parts of tropical Africa. In general, soils of high-activity clays are less easily dispersed than those of low-activity clays.

Soils of the tropics in relation to their agricultural potential have been described by Obeng (1978) in West Africa, Sanchez and Salinas (1981) in tropical America, in the semiarid tropics by Dregne (1983), and by ACIAR (1984). In West Africa soils of the rainforest and semideciduous rainforest within the rainfall regime of 1270 to 2000 mm are generally gravelly, acidic, and of medium to low inherent fertility. About 130 million hectares of forested soils in West Africa have *in situ*-developed concretionary or ironpans (Obeng, 1978). These soils are cultivated to a wide variety of cash and food crops. The savanna soils of regions with 650 to 1400 mm rainfall are mostly gravelly with *in situ*-developed ironpan occurring at shallow depths. In the West African savanna about 113 million hectares are covered by soils with *in situ*-developed ironpans (Obeng, 1978).

The soils of the savanna region of tropical America and of the Amazon basin are mostly Oxisols and Ultisols (Sanchez and Salinas, 1981). These are soils with low pH and are generally deficit in major plant nutrients, especially P. Soils of tropical Australia are similar to those of tropical Asia and West Africa with identical rainfall regime. Isbell (1978) reported that most common soils in the tropics and subtropics of Australia receiving more than 500 mm of annual rainfall are Alfisols, Entisols, Inceptisols, Ultisols, Vertisols, and Oxisols.

A. Soil-Related Constraints to Intensive Arable Land Use

General soil constraints to intensive utilization of uplands are shown in Table 6. In the rainforest and semideciduous forest zones with soils of low pH, the major constraints are low inherent fertility and nutritional imbalance. Although soil chemical problems are overwhelming, crop growth and yields are also limited by soil physical factors once the nutritional constraints are alleviated. The major soil physical constraints are compaction, erosion, and drought. In the semiarid savanna, soils are prone to severe physical problems, the major limitations being compaction, crusting and surface seal, high soil temperatures, low water-holding capacity, and drought. Nevertheless, nutritional constraints also limit food production, especially when soil physical problems have been overcome.

B. Soil Properties

Most soils supporting tropical rainforests are old, with deep weathering from which the clay has been eluviated to the deep horizon. Consequently, the texture of the surface horizons of strongly weathered Oxisols and Ultisols is generally sandy to loamy sand (Table 7). This is not to say that all upland soils that support forest vegetation are coarse textured. Heavy-textured soils occur along the floodplains of large river systems and in regions of volcanic ash-derived soils. Under the forest cover, soil bulk density is generally low and soils posssess a favorable structure. The good soil structure is caused by the flocculation/cementation of clay by organomineral complexes to form stable aggregates. The stable microaggregates are generally within the size range of silt to sand. However, soil structure easily breaks down once the forest vegetation is removed.

Soils in regions with prolonged dry seasons are desiccated up to about 1 m depth. These soils are thus subjected to high temperature extremes

Table 6. General soil-related constraints to intensive utilization of upland soils in the tropics

Soil-related constraint	*Oxisols*[a]	*Ultisols*[a]	*Alfisols*[a]	*Inceptisols*[a]	*Vertisols*[a]
Erosion	2	2	3	3	3
Compaction	3	3	3	2	2
Crusting	1	1	3	3	2
Drought	2	2	3	1	2
Shallow effective rooting depth	3	3	3	1	2
Trafficability	1	1	1	1	3
Supraoptimal soil temperatures	2	2	3	3	2
Chemical/nutritional	3	3	2	1	2

[a]3, Severe; 2, moderate; 1, slight.

Table 7. Physical properties of the surface horizons of major soils of the tropics

Soil	*Region/country*	*Sand (%)*	*Bulk density ($g\ cm^{-3}$)*	*Plant-available water (%)*	*Reference*
Ultisols	Cerrado, Brazil	52	1.23	—	Stone and Silva (1978)
Oxisols	Bahia, Brazil	83	—	—	Silva (1981)
Oxisols	Para Brazil	76	—	—	Baena (1977)
Oxisols	Maranhao, Brazil	35–79	—	—	
Ultisols	Venezuela	84	—	—	Pla Sentis (1977)
Alfisols	Sri Lanka	60	—	—	Panabokke (1958)
Alfisols	North Australia	58	1.49–1.55	—	Mott *et al.* (1979)
Alfisols	Western Nigeria	67–84	1.22–1.55	7–17	Bonsu and Lal (1983)
Alfisols	Northeast Thailand	35–47	1.00–1.25	—	Takahashi *et al.* (1983)
Alfisols	Bahia, Brazil	24–53	—	5–10	Cadima and Alvim (1973)
Alfisols	Hyderabad, India	—	1.55–1.80	2–5	ICRISAT (1978–1979)

when at low moisture content. Under these conditions soils acquire an extremely hard consistency and are often compacted even in their natural state. With the exception of Vertisols and Inceptisols, soils of the semiarid savanna are also characterized by a coarse-textured surface horizon with low organic matter content. These soils are prone to developing surface seals and crust formation with low water-transmission properties (Mott *et al.,* 1979). An example of the formation of surface crust of an Alfisol in western Nigeria is shown in Plate 3.

Structural properties of soils in the tropics are influenced by the organic matter content and the biotic activity of soil fauna. The widely observed phenomenon of microaggregation, especially in subsoil horizons with high clay content, is related to the surface charge properties and to the amounts of oxides and hydrous oxides of iron and aluminum. The highest percentage of the microaggregates or "pseudo-silt" generally occurs in the argillic rather than oxic horizon. Soils containing a high percentage of pseudo-silt or sand have moisture-retention properties similar to silt and sand.

Most uplands with predominantly low-activity clays have low plant-available water reserves, e.g., generally less than 100 mm in the root zone. The effective rooting depth for seasonal crops is shallow, due either to adverse soil physical properties (Vine *et al.,* 1981; Babalola and Lal, 1977) or to nutrient toxicity and imbalance. Consequently seasonal crops are

Plate 3. Alfisols and other soils containing predominantly low-activity clays are prone to development of slowly permeable surface seal or crust.

prone to periodic drought stress throughout the growing season. Furthermore, most of the plant-available water is released at suction of less than 1 bar and often at suction as low as 0.3 bar. The low plant-available water capacity of tropical soils has been reported by Kowal (1970), Obi (1974), Lal (1979), Amezquita (1981), and Mbagwu *et al.* (1983) for soils of Nigeria. For soils of northern Australia, Arndt *et al.* (1963) and van der Graaff (1965) reported 4.3% to 6.8% available water in various horizons of a Tippera family soil. McCown et al. (1979) reported that three Alfisols in Northern Territory, Australia, respectively, retain 17%, 24%, and 33% of gravimetric water content at field moisture capacity. For Alfisols in Hyderabad, India, the available moisture content ranges from 2.3 to 4.6% (ICRISAT, 1978–1979). In tropical America, Pla Sentis (1977) reported that soils of the Guanipa series in Venezuela retain 7.5% moisture at 0.1 bar suction, 5% at 2 bar, and less than 2% at 15 bar (Figure 4). For soils of Puerto Rico, Wolf and Drosdoff (1976) noted the maximum available water-storage capacity of the upper 30 cm of the soil ranged from 3.6 to 6.0

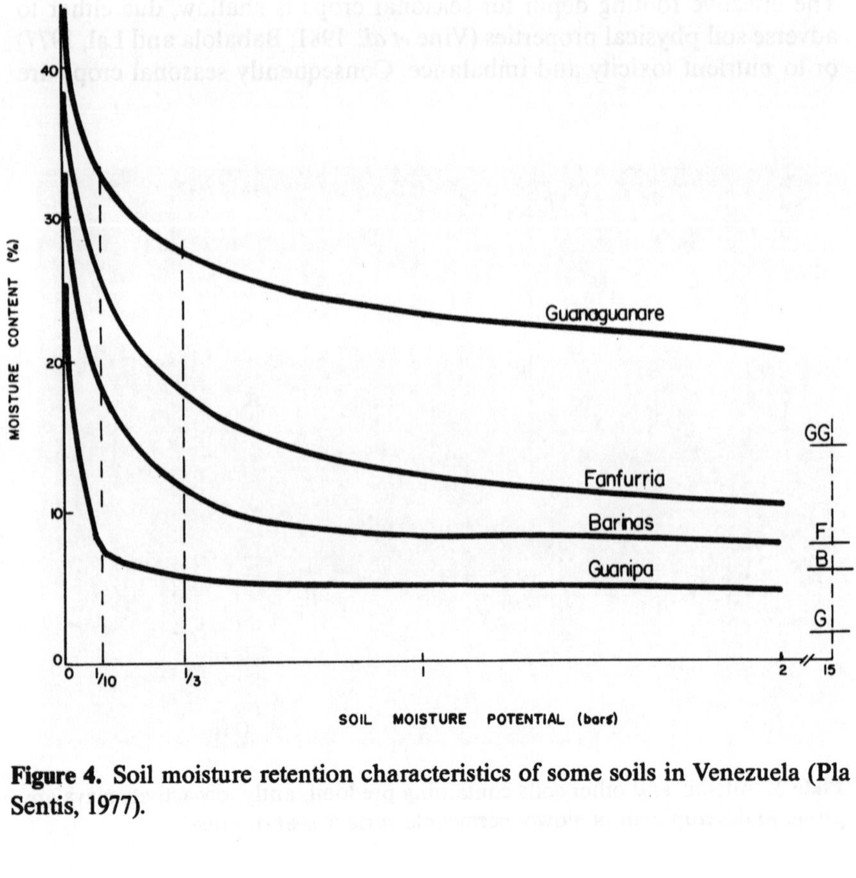

Figure 4. Soil moisture retention characteristics of some soils in Venezuela (Pla Sentis, 1977).

cm but most of the water was released at suctions of less than 1 bar (Figure 5). In the Amazon region of Brazil, Reichart *et al.* (1980) observed that sedentary uplands have low water-holding capacity, measuring approximately 17.7 mm for 0 to 30 cm, 39.4 mm for 0 to 60 cm, and 65.7 mm for 0 to 10 cm depth.

The infiltrability of undisturbed soils of low-activity clays under forest cover is usually very high. The high infiltrability results from both horizontal and vertical components of water flow. The data in Table 8 regarding infiltration measurements made after removing topsoil to varying depths show that the lateral flow is often as deep as about 50 cm in forested lands. The high spatial variability in the infiltration rate of the surface horizon is due to intense biotic activity. In general, the soils in the savanna and semiarid regions have lower infiltration rates than those within forest ecologies. Structurally stable Andosols have also high infiltration rates. The infiltrability of soils of high activity clay, e.g., Vertisols, is low. With cultivation, however, the infiltration rate of most upland declines because of the rapid deterioration in soil structure and the susceptibility of these soils to surface crust and seal.

IV. Traditional Farming Systems and Soil Productivity

Traditional agricultural systems of the tropics, called shifting cultivation and bush fallow rotation, are subsistence, diverse, and based on low

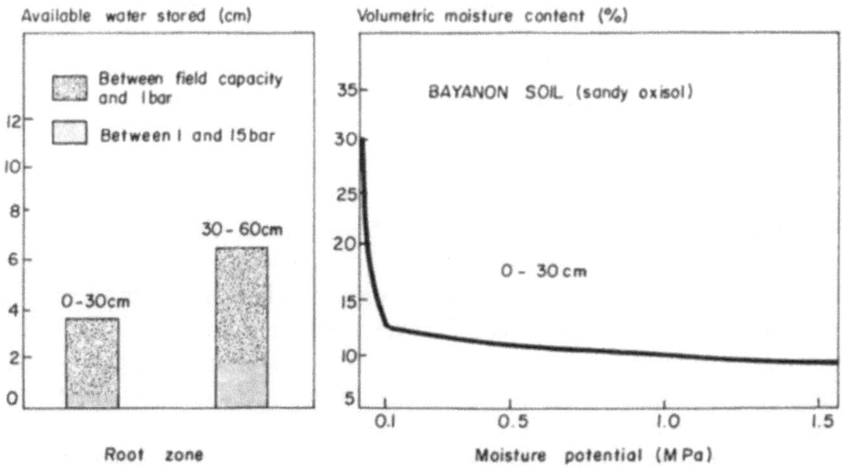

Figure 5. Most of the plant-available soil water is released at suctions of less than 1 bar. Wolf and Drosdoff, *Agro Univ. Puerto Rico,* Vol. 60, pp. 386–394. Copyright 1976 by the University of Puerto Rico.

Table 8. The equilibrium infiltration rate of some Alfisols under forest cover at Ibadan, Nigeria. The infiltration rate was measured by the double-ring infiltrometer after removing 0, 30, and 50 cm of the top soil

	Equilibrium infiltration rate (cm hr^{-1})		
Plot	*0 cm*	*30 cm*	*50 cm*
1	365 ± 283	39 ± 3	39 ± 21
2	159 ± 129	19 ± 12	50 ± 4
3	149 ± 48	55 ± 44	23 ± 8
4	103 ± 77	47 ± 29	50 ± 9
5	54 ± 7	13 ± 4	14 ± 10
6	36 ± 5	22 ± 18	12 ± 3
7	231 ± 21	89 ± 71	11 ± 4
8	76 ± 50	28 ± 21	7 ± 2
9	234 ± 216	43 ± 25	51 ± 17
10	80 ± 3	16 ± 1	10 ± 6
11	323 ± 48	35 ± 1	18 ± 7
12	81 ± 9	18 ± 7	11 ± 7

Unpublished data, R.Cal.

inputs. In some regions, especially on marginal or steep lands of high population density, traditional "slash and burn" farming has been destructive and has caused severe soil and environmental degradation. In most of the lowland tropics with low population density and relatively fertile soils, however, the traditional systems have proved to be ecologically stable despite the minimal inputs made by the shifting cultivators (Greenland, 1975). Most traditional systems of the lowland tropics, subsistence as they may be, have builtin food security and ecological stability.

The magnitude of change in soil properties under traditional systems varies widely among soils, rainfall, vegetation, and the land-use factor, L. The latter, defined as the ratio of cropping period C plus fallow period F to cropping period C [$L = (C + F)/C$], is generally high for shifting cultivation and decreases with more stable and intensive land use. In Casamance, Senegal, Siband (1972) evaluated alterations in soil physical properties of farms under traditional cultivation systems for 2 to 5, 5 to 16, and 50 to 80 years. The cultivation duration caused a marked deterioration of the surface soil with rapid fall in organic mater content, decrease in clay content, and reduction in water-holding capacity and cation-exchange capacity (Table 9). The plots cultivated for 80 consecutive years recorded a decrease of 34% in clay, 47% in silt, 45% in organic matter, 47% in nitrogen, and 17% in the base saturation. Grain

Table 9. Effects of duration of cultivation on soil properties of 0- to 10-cm layer at two village sites in Senegal

Soil property	*Village Sare Bidji (1)*			*Village Diankancounda Oguneul (2)*		
	5 yr	*16 yr*	*80 yr*	*2 yr*	*15 yr*	*50 yr*
Clay (%)	9.1	9.2	6.0	9.3	8.9	6.5
Silt (%)	5.5	4.3	2.9	4.7	4.2	3.3
Sand (%)	85.4	86.5	91.1	86.0	86.9	90.2
Available water (%)	5.7	4.5	4.4	5.6	5.8	5.0
Organic matter (%)	2.4	1.5	1.3	1.5	1.3	1.1
Total nitrogen (%)	0.73	0.54	0.43	0.67	0.49	0.41
pH (1:1 H_2O)	6.5	6.4	6.0	6.6	6.2	5.6
CEC (meq per 100 g)	4.7	3.5	2.6	3.5	3.2	2.3
Base saturation (%)	86	80	69	81	69	52

yield of upland rice was accordingly affected. In Thailand, Takahashi *et al.* (1983) observed that traditional farming resulted in an increase in soil bulk density and soil hardness (Table 10). At International Institute of Tropical Agriculture (IITA), Ibadan, Nigeria, Lal (1984) reported that the rate of change in soil physical properties (e.g., bulk density, infiltration) was less with traditional farming and slash and burn methods than with intensive agriculture (Figure 6; Tables 11a and 11b).

Soil erosion and water runoff under traditional farming also depends on the soil, land-use intensity, relief, and cultural practices adopted. Runoff and erosion are generally less if the fallow period is long enough to restore soil physical properties and build up soil organic matter content. In the Philippines, Kellman (1969) studied runoff and soil erosion during cultivation and fallow phases. He observed that water runoff increased from 1.08% in the first year after clearing to 11.64% in the 12th consecutive year. Lal (1981a) reported negligible runoff and erosion from a traditionally cultivated plot during both the cropping and the fallow phases (Table 12). In Thailand, Takahashi *et al.* (1983) reported that in 1980 soil erosion under shifting cultivation was 81.9% and 56.8% of that under continuous cultivation and bare fallow treatments, respectively. On a sandy savanna soil near Ibadan, Nigeria, Sabel-Khoschella *et al.* (1984) reported that in comparison with the no-till mulch the runoff and erosion on traditionally farmed plots were more on yam planted on heaps than on cassava planted on ridges across the slope. In 1983 erosion on traditional plots was about twice as much on mounded seedbeds as on

Table 10. Effects of shifting cultivation on soil bulk density (g cm^{-3}) near Khon Kaen, Thailand

		Bulk density (g cm^{-3}) at time of observation				
Treatment	*Depth (cm)*	*Jan. 1980*	*Aug. 1980*	*Jan. 1981*	*Oct. 1981*	*Jan. 1982*
Shifting cultivation[a]	0–5	1.148	1.065	1.146	1.045	1.145
	5–15	1.320	1.272	1.307	1.338	1.316
Continuous upland agriculture	0–5	1.148	1.042	1.077	1.082	1.131
	5–15	1.272	1.303	1.345	1.310	1.324
Bare fallow	0–5	1.058	1.031	1.158	1.169	1.247
	5–15	1.242	1.338	1.357	1.398	1.400
Forested control	0–5	0.989	0.910	0.874	0.948	0.988
	5–15	1.245	1.301	1.282	1.214	1.246

[a]Shifting cultivation plot received no-tillage operation. Adapted from Takahashi *et al.*, 1983.

Table 11a. Effects of methods of deforestation on soil bulk density and penetrometer resistance of 0- to 5-cm layer

	Preclearing bulk density ($g\ cm^{-3}$)[a]				*Preclearing penetrometer resistance ($kg\ cm^{-2}$)*[a]			
Treatment	*1978*	*1979*	*1980*	*1981*	*1978*	*1979*	*1980*	*1981*
Traditional farming	0.64	1.06	1.07	1.27	0.21	0.96	0.52	1.32
Manual clearing	0.68	1.17	1.17	1.39	0.20	1.4	0.75	1.19
Shear blade	0.70	1.19	1.37	1.38	0.26	1.0	1.84	2.19
Tree pusher/root rake	0.60	1.24	1.32	1.42	0.20	1.3	0.73	1.23

[a]Each figure is a mean of 25 separate analyses.

Unpublished data of R. Lal.

Table 11b. Effects of land-clearing methods on physical properties of an Alfisol in western Nigeria

Clearing method	*Soil bulk density 0- to 10-cm depth* $(g\ cm^{-3})$ Initial	Final	*Penetrometer resistance* $(kg\ cm^{-2})$ Initial	Final	*Hydraulic conductivity* $(cm\ min^{-1})$ Initial	Final
Mechanical	0.91	1.25	0.50	5.13	16.13	1.29
Slash and burn	0.86	1.12	0.67	2.89	15.23	5.04
Slash	0.89	1.13	0.60	1.86	9.75	4.61
LSD (.05)	0.29		0.35		9.3	

Adapted from Lal and Cummings, 1979.

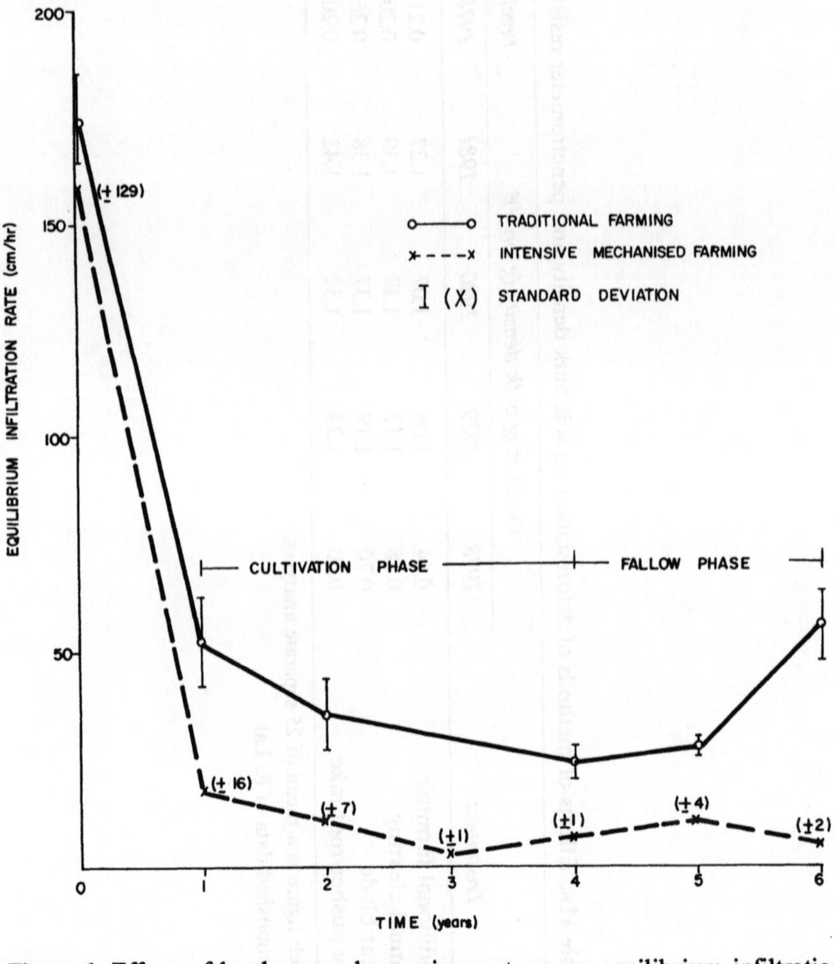

Figure 6. Effects of land use and cropping systems on equilibrium infiltration rate of an Alfisol at IITA, Ibadan, Nigeria.

Table 12. Effects of deforestation on runoff and soil erosion from 1979 to 1981

Method of deforestation	*Runoff (mm)*	*Soil erosion ($t\ ha^{-1}$)*
Forest control	0	0
Traditional	6.6	0.02
Manual	47.7	5.0
Shear blade	104.8	4.8
Tree pusher/root rake	250.3	20.0

Adapted from R. Lal, 1981a.

no-till plots. In 1984 soil erosion was less with ridges made across the slope than with the plowed flat treatment (Table 13). The traditionally hoed plots had higher water infiltration rates than those where farm operations were conducted by motorized equipment (Figure 7). A similar experiment conducted on savanna soil near Ilorin, Nigeria, indicated that erosion on traditionally farmed plots with ridges made up and down the slope was more than on no-till plots but less than on plots that were disk plowed to about 20 cm depth (Table 14).

Traditionally farmers abandon the land when crop yields are too low either because of prevalence of pests (weeds, insects, diseases, etc.) or because of deterioration in soil quality, i.e., compaction, erosion, and decreases in soil organic matter content and plant-available nutrients. The rate of decline in crop yield on soils under traditional farming depends on many factors, e.g., soil properties, crops grown, climate, soil management practices. A study conducted by Allan (1965) in East Africa indicated that on soils of high inherent fertility it took continuous cultivation to cause severe yield decline. On soils of low inherent fertility,

Table 13. Soil erosion, runoff, and cropping systems in a savanna soil

Treatments	*1983* Erosion ($t\ ha^{-1}$)	*1984*[a] Erosion ($t\ ha^{-1}$)	*1984*[a] Runoff mm	*1984*[a] Runoff %
Traditional	10.2	1.5	23.8	1.7
No-till	4.6	0.6	7.2	0.5
Conventional plow	6.8	10.1	154.9	11.4
Bare fallow	33.9	261.6	327.1	24.0

[a]Rainfall 1984 = 1362.3 mm.

Adapted from IITA, 1984.

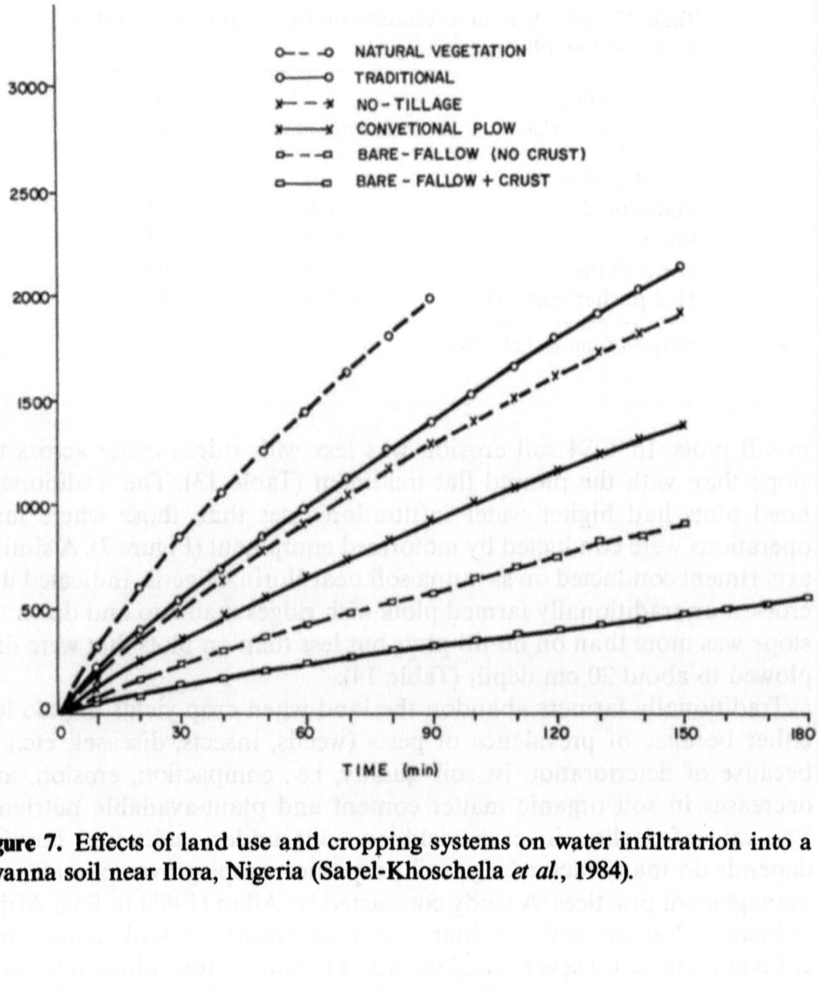

Figure 7. Effects of land use and cropping systems on water infiltratrion into a savanna soil near Ilora, Nigeria (Sabel-Khoschella *et al.*, 1984).

Table 14. Soil erosion and runoff from traditionally farmed, no-till, and conventionally plowed treatments of a savanna soil near Ilorin, Nigeria (1984 data)

Treatment	*Maize grain yield* ($t\ ha^{-1}$)	*Runoff* (*mm*)	*Soil erosion* ($t\ ha^{-1}$)
Traditional farming	2.4	44	1.1
No-till	2.9	36	1.0
No-till with pigeon pea[a] strips	2.3	46	0.2
Plowed	3.6	30	2.4

[a]*Cajanus cajan.*

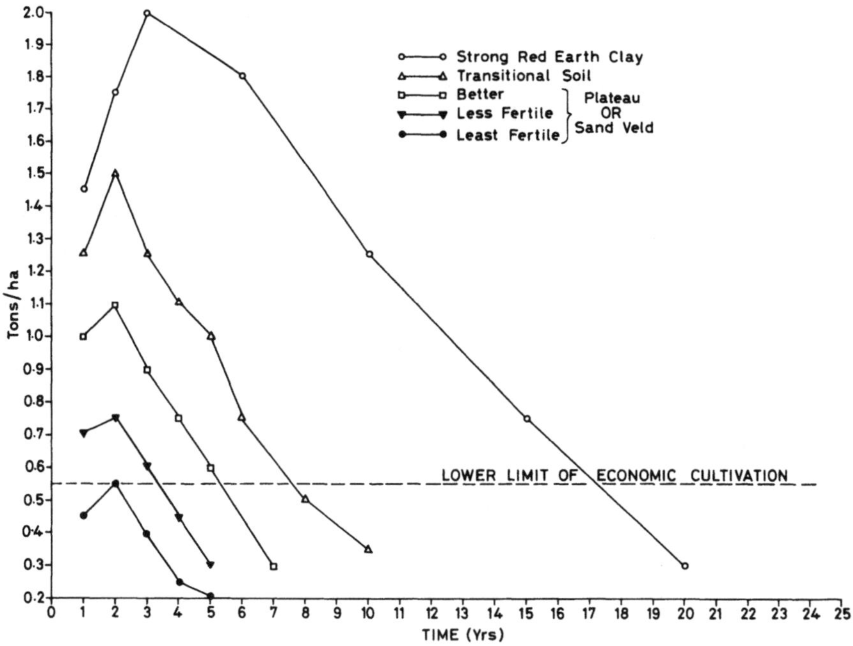

Figure 8. Effects of soil properties and of duration of cultivation of yield decline with traditional farming methods in East Africa (Allan, 1965).

however, yield decline with traditional farming occurred in 1 to 2 years (Figure 8). There are urgent reasons to improve the production per unit time and per unit area even on those soils where yield decline does not occur so rapidly with the traditional shifting cultivation methods. The urgent reaons for improving the traditional farming system is its low productivity. The system is ecologically stable and works as long as those practicing it are willing to remain at the subsistence level.

V. Soil Degradation and Ecological Fragility in the Tropics

Implications of intensive land use in the humid and subhumid tropics in terms of widely reported severe soil degradation have been a severe bottleneck in introducing improved and intensive farming systems. Furthermore, fears of environmental and soil degradation by intensive land use are easily exaggerated because there is a lack of solid data from well-designed, adequately equipped, long-term field studies. Consequently it has become customary to refer to tropical ecology as "fragile" and to the soils as easily "degraded" without precisely understanding the biophysical processes that may or may not lead to the feared effects. The use of word "fragility" in relation to an ecology refers to the ability of an "ecosystem" to resist drastic alterations in its biophysical processes and to revert back to the state that existed before human intervention. In the context of tropical forest and savanna ecosystems the "fragility" refers to irreversible/permanent changes or disruptions in energy balance, in water balance, in cycles of major plant nutrients (C, N, P, K, Ca, Mg, and S), and in major energy pathways. Few studies in the tropics have quantified the effects of alterations in nutrient cycles and energy pathways due to changes in land use.

Similarly, the terms "soil degradation" and "desertification" are rather loosely used without the benefits of quantitative data to define them. The term soil degradation should be defined in terms of the rate of change in soil quality in relation to crop productivity. The soil quality ecompasses organic matter content, biotic activity of soil fauna, soil structure and water infiltrability, porosity and pore-size distribution, cation-exchange capacity, pH, concentration of toxic elements and nutrient imbalance, etc. It is therefore necessary to define the rate of alteration in soil properties affected by change in land use and define their critical limits in relation to crop productivity. For example, before a soil can be dubbed "easily degraded" it is important to define the critical limits of soil organic matter content, porosity, water and nutrient status, and tolerable level of toxic elements beyond which yield of different crops is drastically reduced. Over what time periods under different farming systems and soil and crop management techniques do major soils of the tropics attain these production limits? What are the critical limits beyond which a soil is liable to undergo irreversible degradation?

The terms "desertification" and "degradation" are used interchangeably. Dregne (1978) defined desertification as a process of impoverishment of terrestrial ecosystems under the impact of man leading to (1) reduced productivity of desirable plants, (2) undesirable alterations in the biomass, and in diversity of micro- and macrofauna and flora, (3) accelerated soil erosion, and (4) increased hazard for human occupancy. Once again, the terms "impoverishment," "undesirable alterations," and "hazard" are too vague and must be quantified by defining their critical limits.

A. Processes of Soil Degradation

For soil degradation to be understood and curtailed it is imperative that the cause–effect relationships be known in terms of both the factors and processes of soil degradation. The soil degradation processes influence edaphic factors that result in low crop yields. Some important soil degradation processes are discussed below.

1. Decline in Soil Organic Matter Content

Rapid decline in the soil organic matter content of cultivated soils is a direct effect of continuously high temperatures throughout the year. Some studies have shown that the rate of mineralization of organic matter content in tropical soils may be four times greater than in temperate region soils (Jenkins and Ayanaba, 1977). Consequently cultivated soils in the tropics have lower levels of organic matter than similar soils in the temperate latitude. Lal and Kang (1982) reported large differences in organic carbon status of soils from various ecological regions of Nigeria—forest (1.3 ± 0.08%) > derived savanna (0.98 ± 0.07%) > Guinea savanna (0.7 ± 0.06%). The organic matter content of a soil and its susceptibility to erosion are intimately linked. While a decrease in organic matter content increases susceptibility of the soil to erosion, water erosion also preferrentially removes soil colloids, including the humified organic matter fraction (Lal, 1976). Lal (1981c) reported a linear decline in soil organic matter content with accumulative soil erosion:

$$\text{Organic carbon (\%)} = 1.79 - 0.002E, \; r = -0.71^{**}$$

where E is the annual accumulative soil erosion in tons per hectare. A decrease in organic matter content of the soil also increases its susceptibility to formation of surface crust, which further enhances the risk of soil erosion. Soil erosion is also increased by the reduction in biotic activity of soil fauna that occurs with decrease in soil organic matter content.

In addition to decreases in structural stability, reductions in organic matter content have important implications in terms of plant-available water reserves in the soil. The favorable effects of organic matter content on soil water retention have widely been reported in soils of the tropics and subtropics (Table 15). In fact, organic matter content may have more beneficial effects on the available water-holding capacity than the clay content. Rapid declines in soil organic matter content reduce the soil's structural stability, render soils prone to crusting, increase soil erosion hazards, and decrease plant-available water reserves. Also important are the nutritional implications, e.g., decreases in effective cation exchange capacity (ECEC), acidification, and plant nutrients.

2. Clay Migration

Loss of clay from the surface soil horizon is an important process that leads to soil degradation. The loss of clay content may be caused by the process of eluviation or migration of the clay to subsoil horizons (Roose, 1977). Under forest cover the activity of soil fauna, e.g., earthworms and termites, plays an important role in bringing the fine particles to the soil surface. Nye (1955) stated that termites (*Macrotermes nigeriensis*) transport

Table 15. Effects of soil organic matter content on moisture-retention characteristics and the plant-available water reserves in some soils from the tropics

Region	*Regression equation*[a]	R^2	*Reference*
Field capacity (%)			
Uganda	Y = 7.38 + 0.16 (silt) + 0.30 (clay) + 1.50 (OM)	0.96**	Pidgeon (1972)
Zaire	Y = 3.22 + 0.31 (clay) + 0.35 (silt) + 5.61 (OM)	0.80**	Tran-Vinh-An and Nguba (1971)
Wilting point (%)			
Uganda	Y = 4.19 + 0.19 (silt) + 0.39 (clay) + 0.9 (OM)	0.99**	Pidgeon (1972)
Zaire	Y = 0.25 + 0.32 (clay) + 0.196 (silt) + 2.0 (OM)	0.87**	Tran-Vinh-An and Nguba (1971)
Plant-available water (%)			
Uganda	Y = 169.3 − 1.50 (clay) + 6.09 (OM)	0.89**	Pidgeon (1972)
Surinam	Y = 0.04 (clay) + 0.96 (silt) + 3.4 (OM)	0.78*	Schroo (1976)
Surinam	Y = 2.10 (silt) + 3.3 (OM) + 3.2	0.79*	Schroo (1976)

[a]OM = organic matter. All soil constituents are expressed as percent on weight basis.

Plate 4. A large termite nest in the savanna region of Nigeria.

about 0.025 mm of fine earth fraction to the soil surface annually. Some termites build large nests in tropical savanna and subhumid region, thereby causing a substantial soil turnover (Plate 4). Reductions in soil biological activity decrease the net clay movement to the soil surface, decrease soil porosity, and increase the surface runoff and sediment transport in the overland flow. Accelerated soil erosion leads to reduction in organic matter and clay contents of the surface soil horizons. For example, Lal (1976) reported a significant decrease in clay and an increase in sand and gravel contents of the surface horizon due to accelerated soil erosion (Figure 9).

3. Structural Deterioration and Soil Compaction

An important process leading to soil degradation is the deterioration of the structural properties of a soil and of its ability to regulate water and

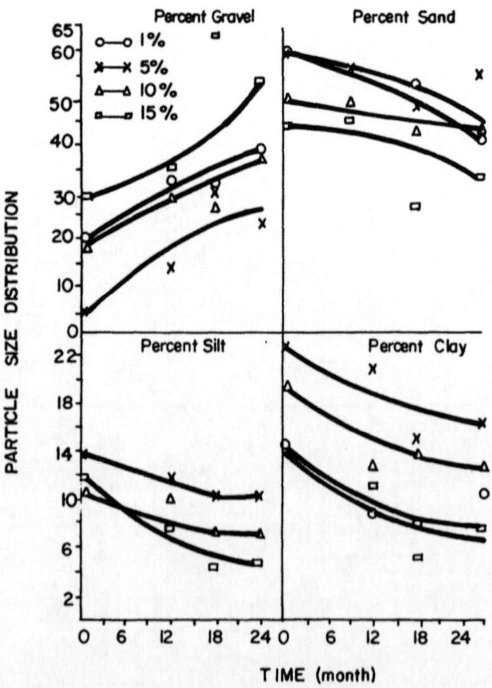

Figure 9. Effects of erosion by water on particle size distribution of the surface soil horizon of an Alfisol on varying slopes (Lal, 1976).

air movement through the profile. Structural degradation, caused by a decline in soil organic matter and clay content and reduction in biotic activity, leads to crusting, compaction, reduced infiltration rate and low available water-holding capacity, increased soil detachability, and accelerated runoff and soil erosion. High erosion risk is a direct consequence of deterioration of soil structure.

Soil compaction is a more severe problem than hitherto anticipated. Soils with low-activity clays have slight or negligible swell/shrink capacity. Declines in soil organic matter content, degradation of soil structure, and drying accompanied by high soil temperatures generally lead to consolidation and compression by mechanisms not well understood. Over and above these factors is the compactive effect of heavy machinery. It is estimated that some 90% of the soil surface may be traversed by tractor wheels during, for example, the primary tillage operations (Soane, 1975). A smearing action of the plow sole results in pore discontinuity that inhibits water movement and root development.

Characterization of soil compaction is another problem that is particularly severe for heterogeneous gravelly soils. The usual criteria, such as bulk density, total porosity, and penetrometer resistance, are not the best indicators of the problem that plant roots experience. Perhaps the void ratio, the specific volume, air : water permeability, or pore-size distribution and continuity may be better indices of plant response than the bulk density. The available research information on these aspects for soils in the tropics is rather scanty. Critical soil bulk density values for root penetration and crop growth are not known for major soils of the tropics. On the basis of the available literature, the data in Table 16 show the critical soil bulk densities that reduce the economic yield of maize (*Zea mays*), upland rice (*Oryza sativa*), sorghum (*Sorghum bicolor*), groundnut (*Arachis hypogea*), cassava (*Manihot esculenta*), and cowpea (*Vigna unguiculata*).

4. Soil Erosion

Soil erosion is an important process responsible for widespread soil degradation and desertification. According to some estimates the topsoil of the world is being depleted at the rate of 0.7% per year (Brown and Wolf, 1984). The FAO (1984) estimated that if soil erosion continues unchecked between 1975 and the year 2000, about 18% of the rain-fed cropland of the developing countries of the tropics will be lost and the rainfed crop productivity will fall by about 29%. Despite its severe implications there is a lack of quantitative data on current rates of erosion in different soils and ecologies, their effects on crop production under different management systems, and on the acceptable levels of soil erosion. The data in Table 17 show that erosion can be severe in some soils and ecologies of the tropics. The magnitude of its adverse effects, however, vary among soils. The solution to the problem of erosion and soil degradation lies in (1) qauntifying erosion rates for different ecologies, (2) establishing cause and effect relationships for various land uses, (3) determining the acceptable level of erosion for different soils and ecologies, (4) determining economic and ecological implications of both on-site and off-site damages, and (5) developing farming systems to prevent soil erosion. Even the effects of soil erosion on crop productivity are not known. The information from a few studies reported in Table 18 show severe yield reductions by erosion. In addition to affecting crop yields, erosion also has severe and adverse off-site effects on natural waters and public property. And yet, the tolerable levels of soil erosion are not known for major soils of the tropics.

5. Nutrient Depletion and Leaching

Unless promptly and adequately replenished, nutrient depletion by crop removal, leaching, runoff and erosion, and volatilization results in the loss of soil fertility. Loss of bases and increases in acidity by leaching are direct results of decline in soil colloids. The leaching losses of nutrients to

Table 16. Critical soil bulk densities for optimal yield of various crops and soil types

Crop	*Critical bulk density* ($g\ cm^{-3}$)	*Soil type*	*Country*	*References*
Maize	1.38–1.43	Alfisol	Nigeria	Couper *et al.* (1982)
	1.40	Alfisol	Senegal	Nicou and Charreau (1980)
Upland rice	1.40	Alfisol	Senegal	Nicou and Charreau (1980)
Sorghum	1.40–1.50	Luvisol	Botswana	Willcocks (1981)
Groundnut	1.40	Alfisol	Senegal	Nicou and Charreau (1980)
Cowpea	1.48	Alfisol	Senegal	Nicou and Charreau (1980)
Cassava	1.38–1.44	Alfisol	Nigeria	Couper *et al.* (1982)
	1.35–1.50	Alfisol	Nigeria	Kayombo and Lal (1984, unpublished)

Table 17. Selected erosion rates in the tropics

Region/ecology	*Criteria*	*Equivalent field erosion rates (ton ha^{-1} yr^{-1})*	*Reference*
A. Africa			
Ethiopia	Sediment load	165	Virgo and Munro (1978)
Tanzania	Bare soil	38–93	Ngatunga *et al.* (1984)
Ivory Coast	Bare soil	138	Roose (1975)
Ghana	Bare soil	100–313	Bonsu and Obeng (1979)
Nigeria	Bare soil	230	Lal (1976)
Lesotho	Sediment load	180	Chakela (1981)
B. Asia			
India	Crop land,	4–43	Dhruva Narayana and Sastry (1983)
	gullies	33–80	
Bangladesh	50% slope	520	Islam (1983)
Java, Indonesia	Imperata	345	Suwardjo *et al.* (1975)
C. Tropical America and the Caribbean			
Trinidad	10–20° slope, bare	490	Ahmad and Brechner (1974)
Northeast Brazil	Cropped land	115	Ramos and Marinho (1980)
El Salvado	Steeplands	130–260	Wall (1981)
Guatemala	Steeplands cultivated to maize	200–3600	Arledge (1980)
Peru	Bare soil	148	Felipe-Morales *et al.* (1978)
Columbia	Crop land	21.5	Rivas (1983)

Table 18. Soil erosion effects on crop productivity

Soil	*Crop*	*Region/ country*	*Yield decline*	*Reference*
Aridisol	Millet	Burkina Faso	0.32 kg ton^{-1} of soil	Fournier (1963)
Alfisol	Maize	Western Nigeria	23% for 2.5 cm loss	Lal (1976)
Alfisol	Maize	Western Nigeria	0.08–0.26 ton ha^{-1} mm^{-1} soil	Lal (1983b,c)
Ultisol	Maize	Cameroon	50% for 2.5-cm loss	Rehm (1978)
Ultisol	Maize	Southern Nigeria	82 to 100% for 5 to 20-cm loss	Mbagwu *et al.* (1984)
Ultisol	Cowpea	Southern Nigeria	14 to 52% for 5 to 20-cm loss	Mbagwu *et al.* (1984)
Ultisol	Maize	Malaysia	Drastic reduction by 15-cm loss	Huat (1974)
Oxisol	Miscellaneous	Guatemala	90% reduction	Freeman (1980)

Table 19. Estimated conversion rates of forest in the tropics

Type	*America*	*Africa*	*Asia*	*Total*
Closed	0.64%	0.625	0.60%	0.62%
Open	0.59%	0.48%	0.61%	0.52%
Total	0.63%	0.52%	0.60%	0.58%

Adapted from O'Keefe and Kristoferson, 1984.

subsoil are increased when deep-rooted perennials, which would otherwise effectively recycle them, are absent.

B. Factors of Soil Degradation

Factors that induce or set in motion the soil degradative processes are intensive land use necessitated by increasing demographic pressure; harsh climatic conditions, e.g., continuously high temperatures and high-intensity tropical rains; and marginal soils that are shallow, steep, or acidic and devoid of essential plant nutrients. Over and above these, human intervention by deforestation, mechanized tillage, monocropping, and simplification of the ecosystem are major factors responsible for soil degradation.

1. Deforestation

Deforestation and forest conversion, major factors of soil degradation in the tropics (Harrison, 1984), are proceeding at the rate of 0.58% per year (Table 19). In addition, 4.4 million hectares are selectively logged. Some major ecological and environmental alterations introduced by deforestation (Table 20) include energy and water balance, alterations in cycles of major nutrients, and biotic factors. For example, removal of forest vegetation is known to cause significant changes in air and soil temperatures and relative humidity (Figures 10 to 12). The increase in maximum soil temperature of the 0- to 1-cm layer by deforestation may be as much as 20 to 25°C (Lal and Cummings, 1979). In addition, the cleared land is easily compacted. The amount of soil compaction is generally greater from mechanical land clearing than from manual methods (Seubert *et al.,* 1977; Hulugalle *et al.,* 1984). Mechanical clearing also results in scraping of surface soil to windrows, and in loss of nutrient capital out of the ecosystem by fire, by leaching, and in runoff and erosion. Roose (1970) observed in the humid region of Ivory Coast that runoff and erosion increased 50 and 1000 times by deforestation. Lal (1981a) reported that mechanized land clearing increased runoff and soil erosion more than manual clearing methods (Table 12).

Table 20. Alterations in Soil and Microclimatic environments by deforestation and intensive cultivation of tropical soils

Hydrologic cycle			Microclimate		Energy balance		Decrease in			Soil flora and fauna		
Decrease in interception by vegetation	Decrease in the water-transmission and -retention characteristics of the soil	Decrease in water uptake from subsoil below 50-cm depth	Increase in temperature amplitude	Decrease in the mean relative humidity	Increase the fluctuations in soil temperature	Change in the heat capacity of the soil	Organic matter	Base status	Nutrient recycling	Decrease in biological activity of macro- and micro-organisms, notably earthworms	Shift in the vegetation type from broad leaves to grasses and from perennials to annuals	Shift in climatic climax
Increase in evaporation	Increase in surface runoff	Increase in the interflow component	Increase in the incoming radiation reaching soil surface		Change in the phase angle, periodicity, and damping depth							

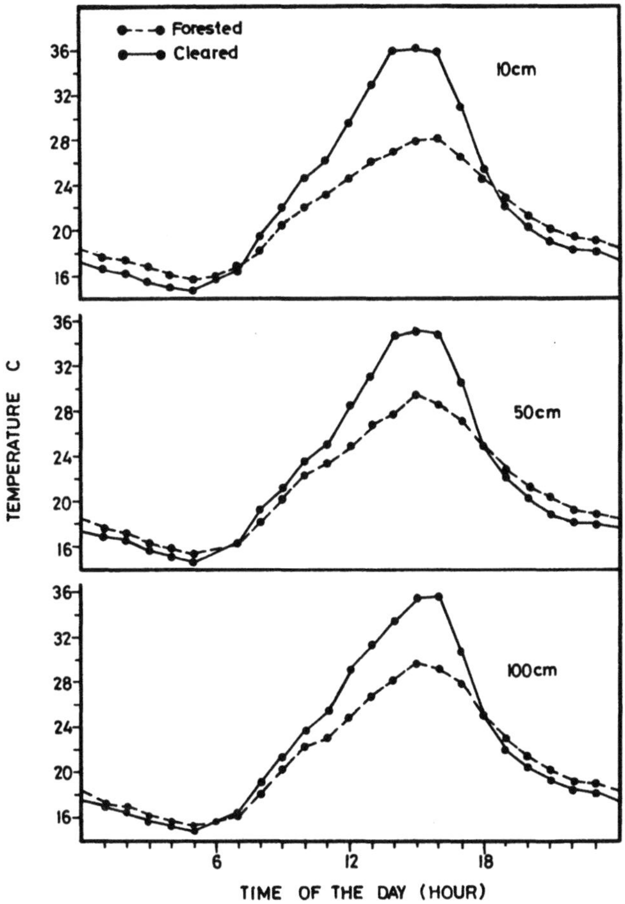

Figure 10. Effects of removal of a semideciduous raniforest at Ibadan, Nigeria on air temperature. Reprinted with permission from Lal and Cummings, *Field Crop Res.*, Vol. 2, pp. 91–107. Copyright 1979 by Elsevier Science Publishers.

2. Motorized Farm Operations and Vehicular Traffic

Soil compaction is inevitable consequence of the introduction of mechanized agriculture in the tropics. It is ironic that theoretical soil mechanics has advanced greatly but its application to practical problems of soil compaction has not been given the attention that it deserves. Silva (1981) reported the effects of mechanized cultivation on the bulk density and porosity of an Oxisol near Bahia, Brazil (Table 21). The mechanized cultivation decreased total porosity of 0 to 30 cm depth from a mean value of 55% to 27%. For an Alfisol in Nigeria Lal (1985) observed significant reductions in water infiltration rate at head points (Figure 13).

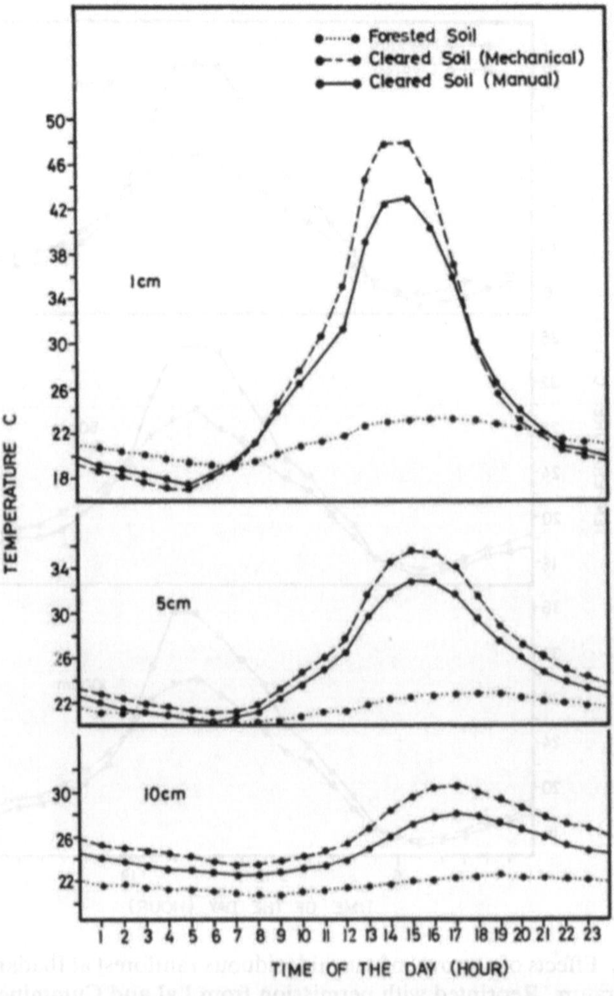

Figure 11. Effects of removal of a semideciduous rainforest at Ibadan, Nigeria on soil temperature. Reprinted with permission Lal and Cummings, *Field Crop Res.*, Vol. 2, pp. 91–107. Copyright 1979 by Elsevier Science Publishers.

The cumulative infiltration after the test began was 28% more within the plot than near the turning point. Maize grain yield for the compacted region was 58% lower than that for the uncompacted soil.

Motorized tillage operation is also a major factor responsible for accelerated soil erosion. Tillage exposes the soil to raindrop impact, enhances the mineralization rate of soil organic matter,and speeds the decline in soil structure. Consequently, erosion and erosion-induced changes in soil properties are more severe on plowed than on no-till soil (Lal, 1983b).

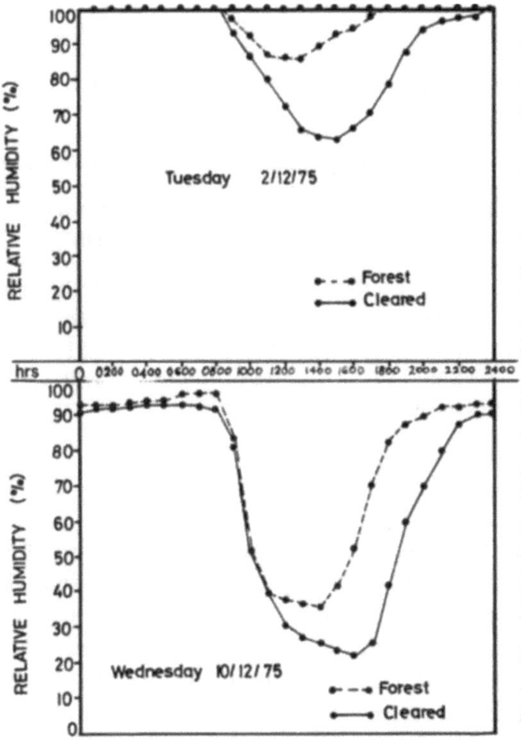

Figure 12. Effects of deforestation on relative humidity (Lal and Cummings, 1979).

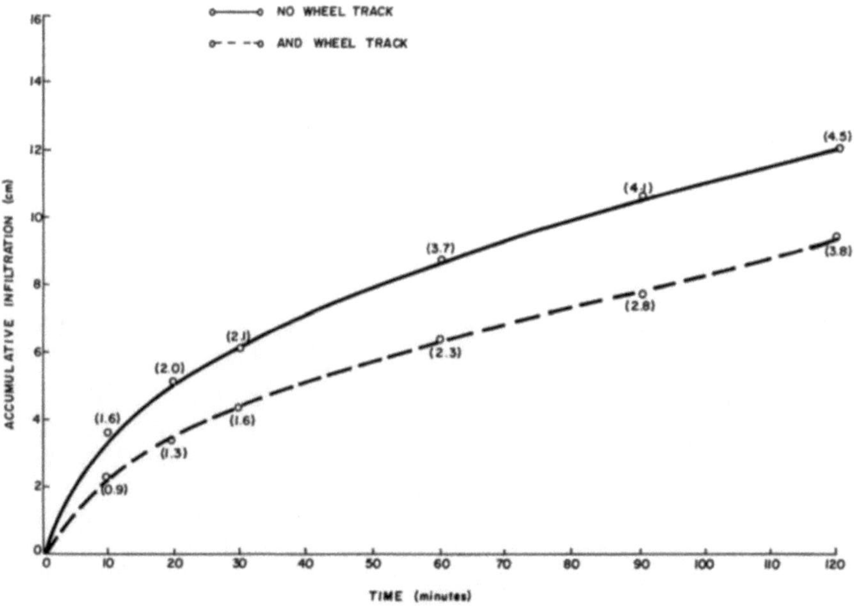

Figure 13. Decrease in water infiltration rate of a tropical Alfisol at the turning points and wheel tracks of the farm machinery (Lal, 1985).

Table 21. Effect of mechanized cultivation on compaction of an Oxisol in Bahia, Brazil (Silva, 1981)

	Without mechanization		*With mechanization*	
Depth (cm)	*Bulk density (g cm^{-3})*	*Porosity (%)*	*Bulk density (g cm^{-3})*	*Porosity (%)*
2.5	1.05	58	1.53	14
10.0	1.29	54	1.64	37
20.0	1.30	54	1.81	22
30.0	1.30	55	1.70	36

3. Fire

Fire is widely used as a management tool in the traditional farming systems of the tropics and subtropics (Plate 5), and thus it is an important ecological factor. Fire is a natural process in some ecosystems (e.g., grass and wood savanna) and the community has evolved to be fire dependent. In contrast the fire-free ecosystems in the tropics (e.g., rainforest, cloud forest, wetlands, swamps, and gallery forests) are greatly perturbed by frequent incidence of human-induced fires. The effects of fire on soil and ecology are more drastic in a fire-free ecosystem. Fire has both short- and

Plate 5. Fire is widely used in traditional farming for land clearing, supplying some plant nutrients, and weed control.

Table 22. Effect of fire on physical properties of soil in Llanos, Venezuela

Soil properties	*Control*	*Burned*
Sand (%)	55.5	57.6
Silt (%)	19.0	16.2
Clay (%)	25.9	26.3
Maximum water-holding capacity (%)	32.8	29.1
Moisture content at −15 bar (%)	7.1	6.5
Field capacity in situ (%)	31.2	25.8
Available soil water (mm)	1850	1327
Evapotranspiration (mm)	1105	1440

Adapted from San Jose and Medina, 1975.

long-term effects. The short-term or immediate effects include a sudden rise in soil and air temperatures, depletion of surface cover by denudation, exposure of soil to climatic elements, changes in soil structure and wettability, alterations in soil texture by intense heat, and destruction of soil fauna. San Jose and Medina (1975) observed a decrease in silt and an increase in sand in a burned soil, and changes in its moisture retention properties and water balance (Table 22). Burning decreases the available water-holding capacity of the soil. Lal and Cummings (1979) observed that soil temperatures can exceed 70°C at 1-cm depth during light burning. After the voluntary savanna burning the denuded soil usually has much higher surface temperatures than the unburned soil protected by vegetation cover. In Madagascar Moureaux (1959) reported the temperatures on bare soil under burned savanna to be 56°, 35°, and 29.5°C for surface, 2.5-cm, and 5-cm depths, respectively. In comparison, the temperatures under unburned land were 40°, 30°, and 26.5°C for surface, 2.5-cm, and 5-cm depths. The long-term effect of repeated fires is the change in climax vegetation from forest to savanna. Large trees of the forest vegetation have given way to fire-resistant shrubs and grasses, resulting in the widespread occurrence of derived savanna in Africa, Asia, and Latin America where rainfall and climatic factors would normally support a forest vegetation. Changes in water balance and loss of soluble bases in savanna soils are partly responsible for the widespread occurrence of plinthite and hardened ironpans (Plate 6).

4. Agrochemicals

Indiscriminate use of chemicals as soil amendments or pest-control measures has important ecological significance because they adversely influence soil biological activities. Soil fauna activity influences soil physical properties and crop residue decomposition. The important effects of faunal activity on soil properties include humification of

Plate 6. The hardened plinthite or ironpan occurs at shallow depths in many savanna soils and severely restricts crop growth.

organic residue, soil mixing and turnover, and the burrowing and tunneling that create macrochannels and water-transmission pores. These activities affect soil aggregation and porosity, water movement and gaseous exchange, weathering and soil formation, erosion control, and nutrient availability. The fauna are confined to the top few centimeters of the soil, the zone that also receives most agricultural chemicals and plant growth regulators. Alterations of temperature and moisture regimens and perturbations of habitat by agricultural practices (e.g., deforestation, tillage, fire) influence soil faunal activity. Agricultural practices also influence their food diversity and availability.

The most drastic effects on faunal activity are those of insecticides and pesticides. In western Nigeria Critchley *et al.* (1979) and Perfect *et al.* (1981) noted that crop protection with DDT reduced the activity and population of soil animals. In watershed management study at IITA, Nigeria, this author observed that merely two applications of Furadon at 2 kg/ha to two consecutive crops to control stem borer in maize eliminated earthworms, termites, and other soil fauna. The soil structure collapsed, compaction became a serious problem soil erosion increased, and crop yield declined. As a result of structural collapse due to elimination of biotic activity, the infiltration rate progressively declined by several orders of magnitude (Figure 14). The application of am-

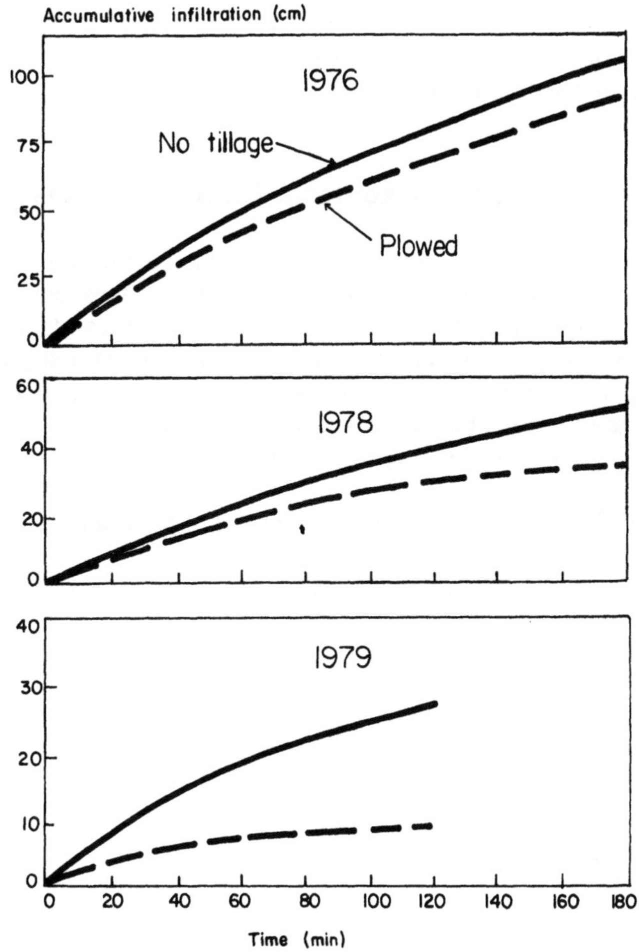

Figure 14. Effects of tillage methods and of soil application of insecticide on cumulative water infiltration (Lal, 1985).

monium sulfate, as a nitrogenous fertilizer, also has adverse effects on soil fauna, notably earthworms. In fact, chemical control of pests has mixed blessings and its use needs to be reviewed in terms of its ecological implications and its effects on soil and environments.

5. Monoculture

The stability of an ecosystem is related to its diversity. The most stable systems (e.g., rainforests or traditional farming systems) are diverse systems. Simplication of a community decreases its stability. The problems of soil erosion and compaction, nutrient depletion, and shifts in floral and faunal species occur after introduction of monoculture over large tracts of land. Even the soil erosion is greater on monocropped than on mixed cropped land (Aina *et al.,* 1977).

VI. Basis of Improving Traditional Farming Systems

Basic principles of soil and resource management are similar for different ecologies, although the technological packages based on these principles vary depending on locally specific constraints and socio-economic considerations. For achieving high and sustained production, while preserving ecological stability and resource base, it is imperative that change in land use and forest conversion include:

1. Preserving the delicate ecological balance among vegetation–climate–soil
2. Maintaining a regular, adequate supply of organic matter on the soil surface
3. Enhancing soil faunal activity and soil turnover by natural processes
4. Maintaining physical condition of the soil suited to the land use
5. Replenishing nutrients removed by plants and animals
6. Creating a desirable nutrient balance and soil reaction
7. Preventing buildup of pests and undesirable plants
8. Adapting a natural nutrient recycling mechanism to avoid leaching losses of nutrients
9. Preserving ecological diversity

The shifting cultivator does all this, which is why the traditional system, evolved over the millenia and used throughout the world, has been ecologically compatible. The basic requirements of a stable system listed above are met if the traditional system allows short cropping periods followed by long fallows. The scarcity of arable land, however, has eliminated or drastically shortened the length of fallow period. The change is, therefore, inevitable. To small landholders in the tropics with limited capital resources and scarce availability of commercial inputs, however, any alterations in the traditional system must also give due consideration to the following factors.

A. Inputs

Improvements in shifting cultivation must be brought about by self-sustaining and low-input technology. This does not imply that the use of synthetic fertilizers and other soil amendments is prohibited. However, soil and crop management systems should be so developed as to minimize the need for them. The nutrient requirement can be met partly

Table 23. Energy balance per hectare in United States and Mexican corn production

Components	*U.S.*[a] *(kcal ha*$^{-1}$*)*	*Mexico*[b] *(kcal ha*$^{-1}$
Input	6,532,071 (28)	642,338 (94)
Output	19,148,700	6,901,200
Ratio	2.93	10.74
Corn yield (kg ha^{-1})	5.934	19,844

[a]Mechanized.

[b]Hand labor.

Adapted from Pimental and Pimental, 1979.

by decreasing losses and increasing efficiency, partly by symbiotic nitrogen fixation and recycling organic wastes, making up the difference by adding synthetic chemicals. The overall energy input must be regulated so that output:input ratio is high.

The high-input agriculture currently being practiced in Western Europe and North America leaves much to be desired. For example, agricultural chemicals amount to 2.5 million kcal or 38.9% of the total inputs needed to produce 1 ha of corn in the United States (Table 23). Although its production is low, Third World agriculture, as it now stands, is more energy efficient than Western agriculture (Table 24). However, the net output of traditional agriculture is low. For example Baylis-Smith (1982) reported 10.4 MJ per person per day output for a traditional farm in New Guinea and 1500 MJ output per person per day, about 150 times more, in North American agriculture. Because of leaving vast areas of land under forest/bush fallow for fertility restoration in traditional farming and because of constraints to intensive land use, the production per unit area per unit time and per person is low. The strategy is to increase output per unit input. Sanchez and Salinas (1981) reported that for Oxisols–Ultisols regions in tropical America the amount of fertilizer

Table 24. Energy budget (output/input) in agriculture

	Third World		*Western countries*	
Crop	*Ratio*	*Country*	*Ratio*	*Country*
Wheat	6.25	India	2	United States
Rice	7.7	Philippines	1.35	United States
Corn	16.67	Mexico	2.7	United States
Sorghum	20	Sudan	3.1	France

Adapted/modified from Knorr, 1983.

required to produce 80% of the maximum or optimum yield is considerably lower than the amount required to reach the maximum or the optimum point. By slightly lowering the desired yield levels, the need for chemical inputs can be drastically reduced.

B. Fertility and Soil Constraints

The objective is not to maximize the production but to achieve a high and desirable level of production by causing a minimum degradation of soil quality. The soil quality must be preserved by preventing soil compaction, preventing soil erosion, and minimizing other soil degradative processes. One of the criteria to assess a restorative technique would be maximize yield per unit of soil loss, or per unit decline in soil organic matter content, pH or ECEC. The fertility-mining systems have no place. The simplest way to increase production would be to grow cultivars and species that are naturally adapted to the specific soil constraints, e.g., root crops in acid soils, rice for Fe toxicity, cowpea for Mn toxicity, etc.

C. Social Considerations

To the small land holders of Africa, Asia, and tropical America labor-intensive technology that can be developed and maintained locally with less skilled manpower and limited capital resources may be more attractive than investment in heavy machinery whose effectiveness is uncertain and whose effects are mostly undesirable. The subsistence farmer of Sub-Saharan Africa who risks famine would consider a successful technology to be the one that produces some yield in the worst year rather than the one that produces high yield in the best. An innovation that would bring about drastic changes in planting habit, discourage mixed cropping, and decrease the flexibility of the traditional system is likely to be unacceptable.

D. Gradual Improvement vs Rapid Transformation

For a shifting cultivator who lives at the subsistence level, lasting impact can be achieved by bringing about gradual improvements over a decade or two in his traditional system rather than through sudden transformation by introducing technology for which he is psychologically unprepared, and physically and financially incapable. Once he gains confidence by adopting simple but productive innovations, as has happened in Western Europe and Asia, high-input technology would naturally follow. Do we have time to adopt this process of slow and gradual improvements?

VII. Components of Technology for Improving Traditional Agriculture

A farming system can be defined as an orderly arrangement of interacting components and subsystems put together to alleviate the biophysical and socioeconomic constraints of a farm enterprise. In other words, a farming system is a strategy to deploy various inputs in the form of various components and subsystems to operate a production farm under a specific environment. Farm environments comprise physical resources (soil, climate, and vegetation), economic conditions, logistics, and social customs and traditions. The system strategy chosen is therefore very specific for each farm enterprise. This report reviews components or subsystems that utilize and manage the physical resources, including vegetation, soil, and climate. Some important subsystems or components are discussed below.

A. Ecologically Compatible Land-Clearing Methods

Land clearing and development are generally considered to be the first steps toward an attempt to improve productivity of the traditional farms. However, improper land clearing and development can rapidly degrade soil quality and decrease crop yields. The magnitude of alterations in biophysical environments depends on the method of land clearing employed. Commonly used methods of land clearing include manual, mechanical, chemical, and combinations of these. The mechanical methods vary widely depending on the choice of attachments used: a chain pulled by two tractors, a dozer blade, a shear blade, a tree extractor, a tree pusher, a tree crusher, a root rake, a root plow, etc. The appropriate method to be used depends on soil and climate, vegetation, envisaged land use, soil and crop management systems, and socioeconomic considerations.

Field experiments conducted in tropical America (Van der Weert, 1974; Seubert *et al.,* 1977; F.T.D. Silva and Lourence, 1977), Africa (Lal and Cummings, 1979; Hulugalle *et al.,* 1984), and Asia (Daniel and Kilasingam, 1975) show that the use of heavy machinery for land clearing results in an increase in soil bulk density and decreases in porosity and water infiltration. For example, the data in Table 11 show that the increase in soil bulk density, penetrometer resistance, and saturated hydraulic conductivity was more for mechanical than for slash or slash and burn methods. Increased compaction of the surface layer drastically reduced the equilibrium infiltration rate in comparison with the forested control. There was a decrease of 74, 66, and 45% in equilibrium infiltration rates for mechanical clearing, slash and burn, and slash alone, respectively. Similarly, Hulugalle *et al.* (1984) observed that the

Table 25. Effects of land-clearing methods on soil chemical properties

Land-clearing method	*ECEC (meq per 100 g)*	*Ca^{2+} (meq per 100 g)*	*Mg^{2+} (meq per 100 g)*	*K^+ (meq per 100 g)*
Mechanical	8.2	7.0	1.7	0.08
Slash and burn	14.0	11.3	2.4	0.10
Slash	9.5	7.6	1.6	0.08
LSD (.05)	1.4	1.1	0.4	0.007

bulk densities of the soils cleared by manual, tree pusher, tree pusher/root rake, and shear blade methods were 1.04, 1.15, 1.7, and 1.18 g cm^{-3}, respectively. Soil bulk density under uncleared forest was 1.00 g cm^{-3}. The increase in bulk density by mechanical methods was due to a reduction in the proportion of large pores (>14.3 μm). Consequently, the saturated hydraulic conductivity was measured to be 0.41, 0.38, 0.26, 0.26, and 0.28 mm sec^{-1} for forest control, manual, tree pusher, tree pusher/root rake, and shear blade treatments, respectively. From their studies in tropical America, Sanchez and Salinas (1981) observed that infiltration rates were 26, 10, and 0.5 cm hr^{-1} in undisturbed forest, slash and burn, and bulldozed treatments at Yurimaguas, Peru. At Manuas, Brazil, the infiltration rate reportedly decreased from 15 cm hr^{-1} under forest to 0.4 cm hr^{-1} under grazed pasture. In Barrolandia, Brazil, the infiltration rates were 24, 20, and 3 cm hr^{-1} in unforested control, slash and burn, and bulldozed clearing treatments.

The effects of land-clearing methods are also reflected in the soil's susceptibility to erosion. Obviously, water runoff and accelerated soil erosion are more severe on mechanically than on manually cleared land. There are, however, differences in erosion hazard among mechanical clearing methods. Methods that leave roots and stumps in the soil, do not remove leaf litter, and cause the least soil disturbance cause less soil erosion. The data in Table 12 show that soil erosion was greater on land cleared with a tree pusher/root rake than where the shear blade land-clearing method was used. In addition to affecting soil physical properties, methods of land clearing also influence chemical and nutritional properties. In general, the soil that has experienced *in situ* burning of the biomass and cut vegetation has more favorable nutritional properties than bulldozed land (Table 25; Sanchez and Salinas, 1981).

Complete versus partial clearing is also a consideration. Researchers in Congo Basin, Zaire, advocated a corridor system of land development.

This involved clearing a strip of forest for food crop production while keeping an alternate strip under forest for soil and water conservation (Jurion and Henry, 1969). In Brazil, Silva (1978) observed no adverse effects form selective logging of commercial timber to clearing, and additional income was provided.

In addition to affecting soil properties, land-clearing methods also affect crop growth and yield. The magnitude of the effect, however, depends on soil characteristics, crop grown, and soil and crop management adopted. Burning has a beneficial effect on crop growth on soils with low pH, but even a slight scraping of the topsoil has drastically adverse effects on crop yield. Seubert *et al.* (1977), from their work in Peru, and Couper *et al.* (1981) in Nigeria observed that some methods of mechanical land clearance may be detrimental to the yield of annual crops grown immediately after the clearance operation. Lal (1981b) evaluated the performance of maize with and without fertilizer and with and without tillage for plots cleared with a front-mounted root rake on a tracked tractor, cleared manually followed by in situ burning, and cleared manually followed by the removal of entire biomass without burning. Maize growth was generally poor on mechanically cleared unfertilized plots. Maize grain yield was generally higher on manually cleared treatments (Table 26). Soil chemical analyses showed that the mean pH was 5.95, 6.0, and 6.2 for mechanical, slash and burn, and slash treatments, respectively. Land clearance treatments also had a significant effect on cation exchange capacity (CEC) and exchangeable cations (Table 25). In a followup study, Hulugalle *et al.* (1984) reported that mechanical methods of land clearing decreased maize grain yield by 6.1, 7.9, and 9.7% for the shear blade, tree pusher, and tree pusher/root rake clearing methods, respectively (Table 27). The adverse effects of mechanical clearing on maize grain yield were attributed to inhibited root development. The yield of deep-rooted cowpea, which followed after maize, was not affected.

The first international symposium on land clearing and development held at IITA in November 1982 endorsed the following guidelines for clearing, development, and protection of tropical lands

1. Increase production of areas now farmed, thereby reducing the need to clear new lands for farming.
2. Halt clearing of steep slopes and inferior soils.
3. Clear manually where possible.
4. Grow cover crops that protect the soil.

The recommendation of manual land clearing followed by in situ burning can be adopted for small landholders where family labor is available or additional labor can be hired at economical rates. Burning

Table 26. Effects of land-clearing methods on maize grain yield

Land-clearing method	*Maize grain yield (t ha^{-1})*		
	First season	*Second season*	*Third season*
Mechanical	4.7	1.4	2.9
Slash and burn	5.1	1.9	4.5
Slash	4.8	1.7	3.7
LSD (.05)	0.7	0.7	1.0

and manual clearing is recommended for savanna regions with sparse vegetation. These regions can be developed either for grain crop production or for pastures. Pasture development from savanna lands in the Northern Territory of Australia is done by chemically poisoning the trees and seeding the grass with improved pasture species without any further soil preparation. Killed trees are left undisturbed. In Cerrado, Brazil, Duque *et al.* (1980) recommended land-clearing methods according to different physiognomic types. For Campo limpo (treeless savanna) and Campo sujo (scattered bush savanna) in situ burning was recommended followed by manual slashing of whatever tree stumps remain. For tree savanna, on the other hand, it is advisable to fell the trees by chaining prior to buring in windrows.

In spite of the established ecological benefits of manual clearing methods, the practice is likely to be less attractive to farmers and organizations interested in developing large areas for mechanized commercial farming. Under these conditions the manual land-clearing operations are slow and inefficient. The use of motorized equipment is thus inevitable. Land-clearing techniques and post clearing soil and crop management practices should be so adapted as to minimize the adverse effects of mechanized clearing. If mechanized clearing must be employed, care should be taken not to (1) remove the litter, roots, or stumps; (2) scrape off or disturb the topsoil; (3) clear the land when the soil is wet; (4) compact the subsoil with heavy machinary, and (5) drag trees or stumps over long distances. The effects of postclearing soil and crop management practices on soil quality and productivity are often more important than that of the one-time land-clearing operations.

B. Seedbed Preparation and Appropriate Tillage Techniques

The objective of seedbed preparation is to facilitate seedling emergence and crop establishment, reduce weed competition, alleviate labor

Table 27. Effects of land-clearing methods on grain yield of maize and cowpea

	Maize		Cowpea	
Clearing method	*t ha⁻¹*	*Relative measure*	*kg ha⁻¹*	*Relative measure*
Manual	4.44	100	390	100
Tree pusher	4.09	92	390	100
Tree pusher/root rake	4.01	90	430	110
Shear blade	4.17	94	430	110
LSD (.05)	0.22		50	

constraints, and conserve soil and water. These objectives can be achieved by a range of techniques, e.g., mechanical, chemical, and biological. Traditionally seedbed preparation has been used synonymously with mechanical soil manipulation performed at a desirable intensity to suppress weed and provide the needed seed–soil contact. A wide range of tillage implements to invert, loosen, mix, level, and pulverize the soil are developed to be human, animal, or machine powered. Tillage operations thus performed affect soil, microclimate, and floral and faunal population. Mechanical tillage operations are capital intensive, (have potentially high risks to) accelerate soil erosion, and increase soil degradation. The most relevant concern of agromomists and environmentalists has therefore been whether mechanical tillage is indispensable to achieving the desired objectives. If needed, how often, and with what equipment?

The economic availability of herbicides and other growth regulators have made it possible to control weeds without mechanical tillage. Weeds can also be controlled by fallowing with aggressive quick-growing cover crops that either die naturally or are easy to control. An important concern, therefore, is under what soil and climate conditions and for which crops can the primary and secondary tillage operations be either completely eliminated or drastically reduced?

1. Soil and Ecological Factors Favoring a No-till System

A no-till or reduced-tillage system is applicable for soils with (1) a coarse-textured surface horizon, (2) good internal drainage, (3) high biological activity of soil fauna, (4) favorable initial soil structure and friable consistency over a wide range of soil moisture, and (5) an adequate quantity of crop residue mulch. These conditions are generally fulfilled for Alfisols, Oxisols, and Ultisols in the lowland humid and subhumid tropics. Under these conditions conventional methods of seedbed preparation involving primary and secondary tillage operations have often been found to be unnecessary for optimum returns. Furthermore, mechanical tillage has been shown to have severely adverse effects on soil as is evident from the review cited below.

Field experiments conducted on Alfisols in western Africa have indicated beneficial effects on soil physical properties from reduced tillage methods (Baffoe-Bonnie and Quansah, 1975; Wilkinson and Aina, 1976; Aina, 1979; Lal, 1983). The data in Table 28 compare the chemical and mechanical properties of the 0 to 5-cm layer of an Alfisol after 24 consecutive crops of maize with no tillage and conventional plowing methods of seedbed preparation. The soil pH in no-till plots was about one-half unit more than that of the plowed treatment and contained 17% more organic matter, 92% more nitrogen, 85% more Ca^{2+}, 3.13 times more Mg^{2+}, and 3.31 times more K^+. The CEC was almost double that in the plowed soil.

High soil organic matter content and favorable soil structure allow the no-till soil to retain more available water in the root zone than soil repeatedly plowed. The data in Figure 15a compare the moisture retention characteristics of the surface soil from plowed and no-tillage treatments. The surface soil from no-tillage treatments retained more water. During periods of drought stress, crops in plowed plots wilted 2 to 3 days earlier than those in no-tillage plots.

In addition, plowed soil is very susceptible to surface sealing. The impacting rain drops from an intense rain break up unprotected structural aggregates and the surface soil develops a semiimpermeable

Table 28. Effects of tillage methods on properties of the 0 to 5-cm soil layer of an Alfisol after growing 24 consecutive crops of maize

Property	*No tillage*	*Conventionally plowed*
Sand (%)	68.9 ± 2.5	70.9 ± 1.9
Silt (%)	11.2 ± 1.6	9.9 ± 0.9
Clay (%)	19.9 ± 1.9	19.2 ± 1.6
Gravels (%)	12.1 ± 2.5	27.5 ± 9.8
pH (1:1 in H_2O)	5.9 ± 0.1	5.3 ± 0.2
Organic carbon (%)	1.69 ± 0.27	1.44 ± 0.07
Total nitrogen (%)	0.179 ± 0.040	0.093 ± 0.007
CEC (meq per 100 g)	7.84 ± 1.40	3.99 ± 0.91
Exchangeable Ca^{2+} (meq per 100 g)	5.31 ± 0.84	2.87 ± 0.62
Exchangeable Mg^{2+} (meq per 100 g)	0.97 ± 0.21	0.31 ± 0.08
Exchangeable K^+ (meq per 100 g)	1.06 ± 0.46	0.32 ± 0.16

crust that impedes water and air movement. The development of surface crust restricts water entry into the soil. The data on infiltration (Figure 15b) show that the equilibrium infiltration rate was 18 and 48 cm hr^{-1} for plowed and no-tillage soil, respectively. As a result, water runoff and erosion are considerably greater from plowed soil. The data in Table 29 on runoff and soil erosion from field plots at IITA for 1973 indicate that the no-tillage treatment prevented runoff and soil erosion up to 15% slope. For example, in the first season 1973 the runoff from no-till maize was 2.5, 1.1, 38.7, and 23.4% of that from plowed maize for 1, 5, 10, and 15% slopes, respectively. The soil erosion from no-till plots was practically negligible. These differences are reflected in the soil loss–slope relationships for plowed and no-till treatments shown in Table 30. Both the coefficients and exponents of the slope variable *S* are more for plowed plots. However, the effectiveness of no-till in reducing runoff and soil erosion depends on having an adequate quantity of crop residue mulch. If the residue from the previous crop is as adequate as it is after growing maize, for example, no-till cultivation is as effective an erosion-control measure as 6 t ha^{-1} of brought-in mulch (Table 31, Figure 16).

The efficiency of applied fertilizer also varies among tillage systems. A no-tillage system containing mulch material of a high C : N ratio may exhibit chlorotic symptoms of nutrient deficiency for zero or low rates of N application during the first one or two seasons after adopting the no-till system. However, when the immobilization and release of nitrogen have

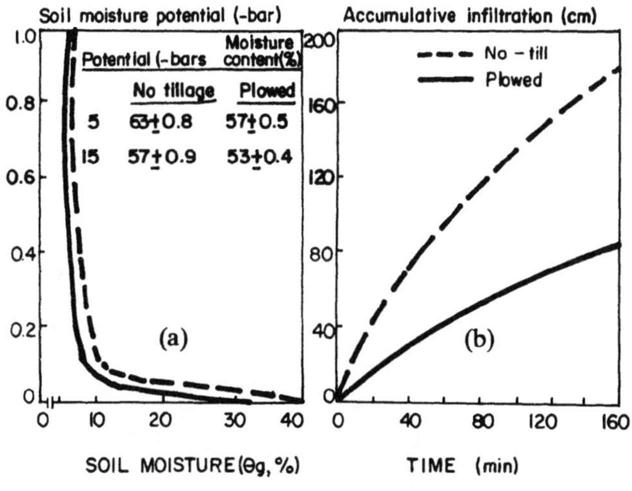

Figure 15. Effects of no-till and plowed methods of seedbed preparation on (a) soil moisture retention characteristics, and (b) water infiltration into an Alfisol at IITA, Ibadan, Nigeria.

Table 29. Effects of slope and soil management on runoff (mm) and soil loss (t ha^{-1}) in 1973; rainfall: first season, 781 mm; second season, 416 mm

Treatment	*Water loss from slopes of:*				*Soil loss from slopes of:*			
	1%	*5%*	*10%*	*15%*	*1%*	*5%*	*10%*	*15%*
(a) First season 1973								
1 Bare fallow	315.7	347.3	311.0	316.5	7.5	80.4	152.9	155.3
2 Maize (mulched)	0.0	6.9	20.3	16.8	0.0	0.0	0.1	0.0
3 Maize (plowed)	55.7	158.7	52.4	89.9	1.2	8.2	4.4	23.6
4 Maize (no tillage)	1.4	1.8	20.3	21.0	0.0	0.2	0.1	0.1
5 Cowpea (plowed)	19.8	81.2	51.4	46.1	0.6	5.6	3.2	7.6
(b) Second season 1973								
1 Bare fallow	191.7	195.8	193.1	185.4	3.7	75.8	79.7	73.9
2 Maize (mulched)	0.0	4.0	9.0	7.5	0.0	0.0	0.1	0.0
3 Maize (plowed)	9.1	65.0	36.0	71.4	0.4	2.8	2.8	17.1
4 Cowpea (no tillage)	6.0	6.5	10.2	10.7	0.0	0.0	0.1	0.0
5 Maize (plowed)	22.2	64.5	30.5	105.4	0.3	4.0	3.0	35.4

Adapted/modified from Lal, 1976.

reached a steady state and losses due to runoff and erosion are contained, the fertilizer use efficiency is generally greater on untilled soil (Figure 17). Experiments conducted at IITA showed that nitrogen-use efficiency by maize was better on no-till than on plowed soil after 6 years of no-till farming (Table 32a). On a lateritic soil in Thailand Tawonmas *et al.* (1984) reported equal or more yields with no-till than with conventional tillage treatments (Table 32a).

The beneficial effects of no-till on soil and water conservation and management have also been reported for other soils by Kannegieter (1969), Ajunwon *et al.* (1978), Juo and Lal (1978), Agboola (1981), Olaniyan (1983) in Africa, by Melville by Melville (1978), and McCown *et al.* (1985) in northern Australia, and by Wijewardene (1981) in Sri Lanka. The beneficial effects of no-till mulch for an Ultisol in Peru have also been reported. Sanchez and Salinas (1981) reported from their trials in Yurimaguas, Peru, that for the first planting after land clearing there is little need for tillage. They observed no significant difference in upland rice yield between no-till and rototill methods of seedbed preparation. Establishment of pasture species in cleared land was also better with no-till. On a Brazilian Oxisol, Sidiras *et al.* (1982) observed significant improvements in water-stable aggregates and soil water retention at 0.06, 0.33, and 1 bar suctions after 4 years of continuous no-till farming. Improvements in soil moisture retention in the no-till system were more pronounced in the top 20-cm layer than in the soil beneath. In addition to improvements in soil structure and infiltration rates, the presence of crop residue mulch on no-till plots decreased runoff and soil losses from an Oxisol on a 4% slope. In Rio Grande Do Sul, Machado (1976) also

Table 30. Soil loss–slope relations for each season of maize–cowpeas (no-tillage) and cowpeas–maize (plowed) treatments (1973 data only)

	Rainfall > 25 mm		*Rainfall < 25 mm*	
Treatment	*r*	*Equation*	*r*	*Equation*
(a) First season				
Maize–cowpeas (no tillage)	0.10	$E = 0.05S^{0.11}$	0.12	$E = 0.12S^{0.11}$
Cowpeas–maize (plowed)	0.12	$E = 0.11S^{0.20}$	0.17	$E = 0.22S^{0.37}$
(b) Second season				
Maize–cowpeas (no tillage)	0.25	$E = 0.04S^{0.20}$	0.11	$E = 0.10S^{0.07}$
Cowpeas–maize (plowed)	0.74	$E = 0.40S^{0.59}$	0.31	$E = 2.11S^{0.41}$

Adapted from Lal, 1976.

Table 31. Effect of mulch rate on runoff (mm) on different slope[a]

Slope (%)	Mulch rate (t ha^{-1}) 0	2	4	6	No tillage
(a) First-season runoff (mm) (April–July)					
1	283	6	4	0	6
5	346	61	10	7	9
10	219	46	21	12	15
15	294	47	20	12	14
(b) Second-season runoff (mm) (Aug.–Nov.)					
1	129	30	3	0	5
5	137	65	18	4	6
10	84	28	14	9	9
15	80	40	31	8	9

[a]Rainfall first season 510 mm, second season 249 mm.
Adapted from Lal, 1976.

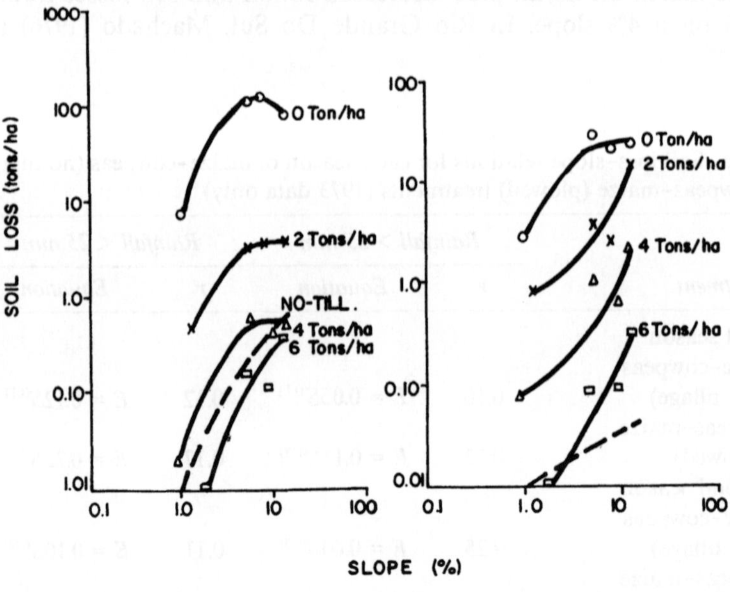

Figure 16. Effects of mulch rate on soil loss from field runoff plots on different slopes (a) first season 1974, (b) second season, 1974.

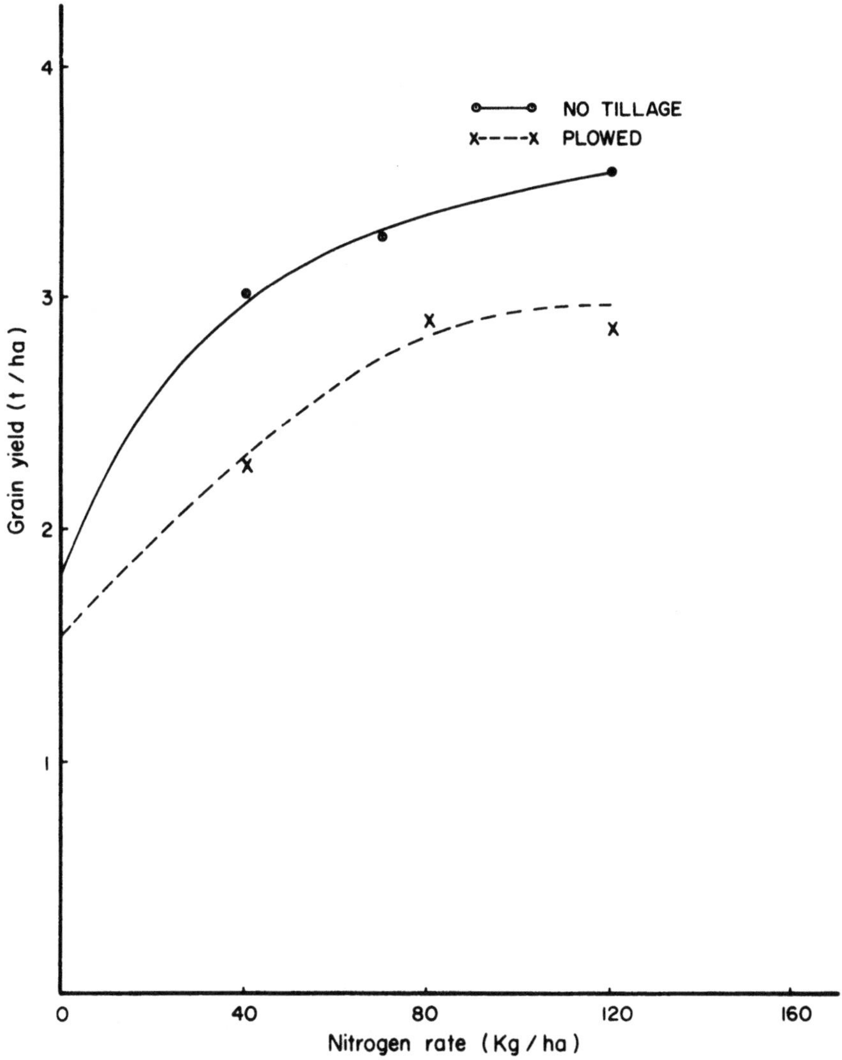

Figure 17. Nitrogen response of maize for two tillage systems for an Alfisol (Lal, 1983b).

reported more favorable levels of soil organic matter content and physical properties with no-till than with conventional plowing (Table 33).

In addition to benefits in soil and water conservation, the no-till system also optimizes the soil temperature regime. Crops grown with the plowing method of seedbed preparation suffer from supraoptimal soil temperatures (Lal, 1983; McCown *et al.*, 1985). The maximum soil temperature in no-till soil is often 8 to 16°C lower than that of a plowed or

Table 32a. Nitrogen response of maize at different levels of P for no-till and plowed systems of seedbed preparation

P rate (*kg ha^{-1}*)	*Nitrogen response functions*[a]	*Correlation coefficient* (*r*)
A. Main season		
(a) No tillage		
0	$Y = 1.06 + 0.015X - 0.00003X^2$	0.93
30	$Y = 1.11 + 0.33X - 0.00016X^2$	0.98
60	$Y = 1.38 + 0.015X - 0.00008X^2$	0.98
(b) Plowed		
0	$Y = 0.69 + 0.024X - 0.00012X^2$	0.99
30	$Y = 0.88 + 0.012X - 0.00005X^2$	0.99
60	$Y = 0.83 + 0.029X - 0.00014X^2$	0.93
B. Short season		
(a) No tillage		
0	$Y = 2.23 + 0.029X - 0.00009X^2$	0.97
30	$Y = 1.70 + 0.013X - 1.00006X^2$	0.99
60	$Y = 2.56 + 0.028X - 0.0001X^2$	0.71
(b) Plowed		
0	$Y = 1.50 + 0.034X - 0.0014X^2$	0.90
30	$Y = 1.23 + 0.048X - 0.00024X^2$	0.94
60	$Y = 2.21 + 0.023X - 0.0001X^2$	0.94

[a]X = N rate (kg ha^{-1}); Y = maize grain yield (t ha^{-1}).
Unpublished data of Lal.

Table 32b. Effects of different tillage methods on maize grain yield for Pak-Chong lateritic soil at the Phraphuttabat Field Crop Station in Thailand[a]

Tillage method	*Year* 1979	1980	1981	1982	*Mean*
Conventional tillage	3.0	4.3	3.9	2.7	3.5
Fine tillage	2.6	4.5	4.9	2.8	4.0
No tillage	3.1	5.2	4.2	4.0	4.1

[a]Fertilizer: 95 kg ha^{-1} N per year at 625 kg ha^{-1} rock phosphate applied in 1979 and 1981.
Adapted from Tawonmas *et al.*, 1984.

Table 33. Effects of tillage methods on physical properties of 0 to 15-cm depth of a Brazilian soil

Tillage treatment	*Organic matter (%)*	*Bulk density ($g\ cm^{-3}$)*	*Porosity (%)*	
			Total	*Macro*
No tillage	3.4	1.21	54.7	14.3
Conventional tillage	1.5	1.35	49.4	6.9

Adapted from Machado, 1976.

ridged soil. Futhermore prevention of surface seal in no-till soil facilitates free gaseous exchanges between the soil and atmosphere. The oxygen diffusion rate in no-till mulched plots is more than in plowed treatments (Table 34) due to lack of crust formation and high biological activity.

The effectiveness of a no-till system in soil and water conservation is lessened if the seeding and harvesting operations are done by motorized equipment. This is because of the soil compaction effects of vehicular traffic. The infiltration rate progressively declines with time due to cumulative effects of vehicular traffic on soil compaction. The data on changes in infiltration rate with time from Alfisols managed with motorized no-till and plowed systems is shown in Figure 14. The cumulative infiltration 2 hr after the beginning of the test for no-till and plowed watersheds decreased from 77 and 65 cm in 1976 to 38 and 28 cm in 1978, 28 and 9 cm in 1979, and 12 and 5 cm in 1980, respectively. This drastic decline in infiltration capacity indicates structural collapse and elimination of transmission pores. Nevertheless the plowed watershed was more compacted than the no-tillage treatment and was more prone to crusting and surface seal. For example, the infiltration capacity in 1979 was 3.8 cm hr^{-1} in the plowed and 10.4 cm hr^{-1} in the no-till watershed. Despite the soil compaction, runoff and soil erosion were still less from no-till than from the plowed watershed (Table 35). An example of a hydrograph from both watersheds is shown in Figure 18. The respective peak rates of flow were 1.45 and 3.60 cm hr^{-1} for no-till and plowed watersheds.

2. Soil and Ecological Factors that Require Specific Adaptations for No-till and Reduced-Tillage Systems

No-till systems are generally ineffective or are less effective on soils with degraded antecedent soil physical conditions and inadequate residue mulch. It is the crop residue mulch on untilled soil that decreases runoff

Table 34. Effects of tillage methods on oxygen diffusion rate (ODR) of an Alfisol growing maize at IITA, Ibadan, Nigeria

Treatment	*ODR* ($\mu g\ cm^{-2}\ min^{-1}$) *at different growth stages*			
	I	*II*	*III*	*IV*
Unplowed and mulched	0.516a	0.652ab	0.722a	0.682ab
Unplowed and chiseled	0.599a	0.530b	0.674a	0.630ab
Moldboard plowing and harrowing	0.458a	0.729a	0.637ab	0.653ab
Disk plowing and rotovation	0.568a	0.588ab	0.696a	0.719a
Unplowed without mulch	0.441a	0.630ab	0.720a	0.665ab
Moldboard plowing at rain ending	0.521a	0.577ab	0.565b	0.592ab
Moldboard plowing harrowed and mulch	0.568a	0.604ab	0.703a	0.66 ab
Moldboard plowing and riding	0.583a	0.637ab	0.704a	0.530a
LSD (.05)	0.193	0.138	0.099	0.138

Figures followed by similar letters are statistically similar within vertical rows.

Unpublished data of Lal, 1983.

Table 35. Runoff and erosion measurements with H-Flume for 1979 on about 5-ha watersheds managed with mechanized no-till and plowed systems of seedbed preparation

Parameter	No tillage			Conventional tillage		
	First season	*Second season*	*Total*	*First season*	*Second season*	*Total*
1979						
Runoff (mm)	3.60	17.93	21.53	173.64	51.42	225.06
Erosion (t ha^{-1})	0.012	0.118	0.130	4.94	0.59	5.50
Rainfall (mm)	583.5	257.8	841.3	583.5	257.8	841.3
1980						
Runoff (mm)	25.13	9.31	34.44	122.80	30.16	152.96
Erosion (t ha^{-1})	0.307	0.025	0.33	1.61	0.286	1.89
Rainfall (mm)	621	279	900	621	279	900

Adapted from Lal, 1984.

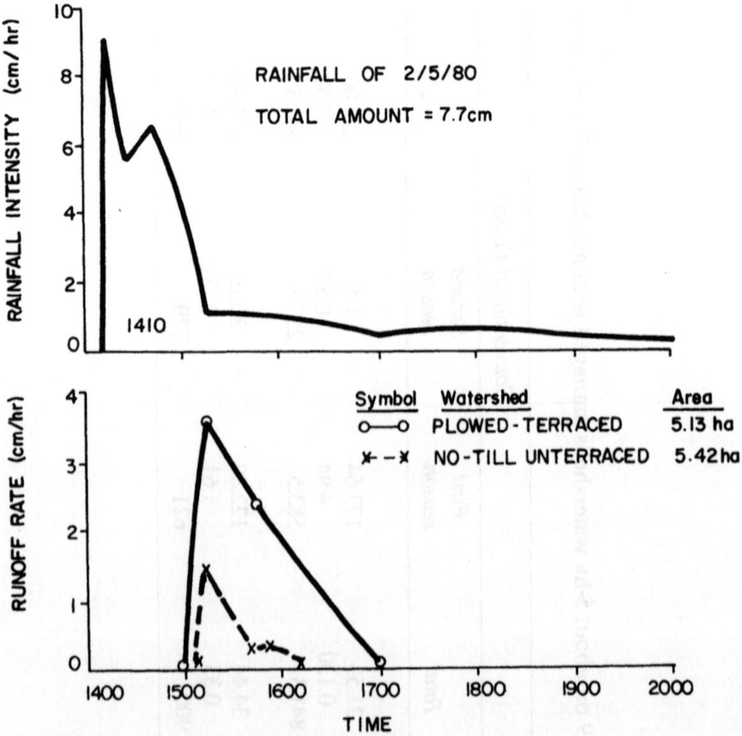

Figure 18. Hydrograph produced during the 2 May 1980 rainstorm on the plowed and terraced and the no-till unterraced watershed managed with mechanized farm operations (Lal, 1984).

and soil erosion (Table 36), regulates soil temperature (Figure 19), and conserves soil water. For example, crop establishment with no-till is generally unsatisfactory on soils that have compacted and crusted surface soil, uneven ground surface, poor seed–soil contact, and inadequate amounts of crop residue mulch. These conditions usually occur in soils of the semiarid tropics, e.g., Ustropepts, Ustalfs, Alfic, Eutrustox, Vertisols, and soils with ironpans. Among them, the soils that contain at least 20% clay with some amount of high-activity clays are "structurally active." Some form of no-till or reduced-tillage system is applicable on them. For example, Dunham and Aremu (1979) reported significantly higher infiltration rates on soils of the semiarid region of Zaria, Nigeria with no-till (54 mm hr^{-1}) than with plowing and harrowing (18mm hr^{-1}). Sturctural stability of aggregates and crop yields were also favorable with no-till or reduced-tillage systems (Dunham, 1982a, b). For others, with structurally inactive surface horizons, some form of mechanical tillage is necessary. Nicou (1977) and Charreau (1977) have shown from their studies in the Sahel, significant increases in plant-available water by deep

Table 36. Slope–runoff relationships for different mulch rates

Mulch rate (t ha)	*Correlation coefficient, r*	*Equation*[a]	*Mean runoff (%)*
(a) Rainstorms > 25 mm			
0	−0.1	$W = 59.2S^{-0.04}$	55.7
2	0.4	$W = 2.1S^{0.66}$	6.7
4	0.76	$W = 0.8S^{1.05}$	2.7
6	0.89	$W = 0.001S^{3.3}$	0.6
No tillage	0.57	$W = 1.20S^{0.29}$	1.9
(b) Rainstorms < 25 mm			
0	−0.27	$W = 47.5S^{0.22}$	33.0
2	0.49	$W = 1.5S^{0.47}$	3.4
4	0.73	$W = 0.8S^{0.66}$	2.3
6	0.90	$W = 0.002S^{3.07}$	0.4
No tillage	0.48	$W = 1.62S^{0.30}$	2.6

[a]W = (Runoff/rainfall) × 100; S = Slope (%)
Adapted from Lal, 1984.

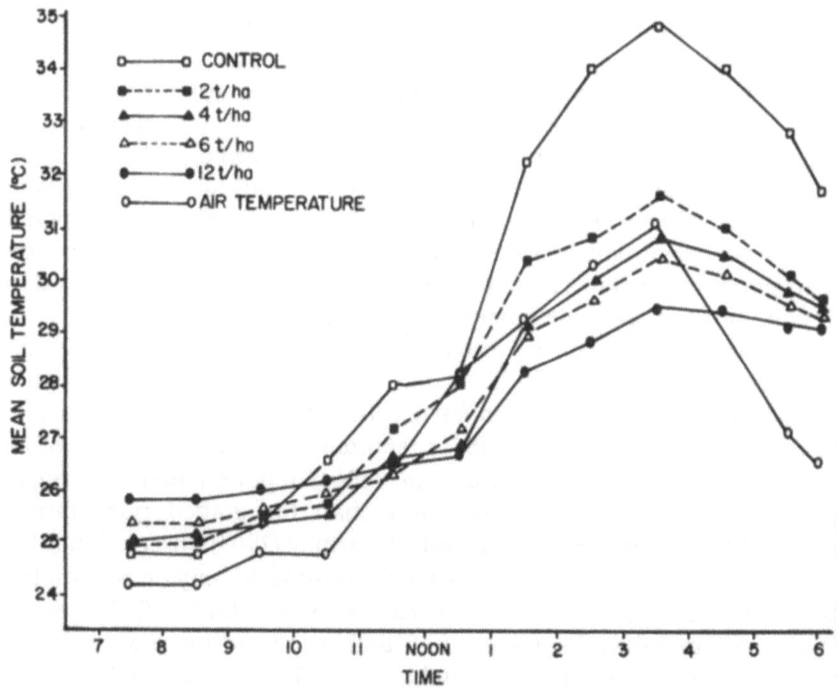

Figure 19. Effects of mulch rate on diurnal fluctuations in soil temperature regime. Reproduced from Lal *et al., Soil Sci. Soc. of Am. J.*, Vol. 44, pp. 827–833, 1980, by permission of Soil Sci. Soc. Am.

tillage and soil inversion due to reduction in losses by water runoff and evaporation. Plowing has been shown to increase total porosity and root growth (Nicou, 1974a). Nicou (1979), Chopart (1981), and Chopart *et al.* (1981), while summarizing the results of tillage studies conducted in Senegal, Togo, and Ivory Coast, concluded that mechanical tillage had beneficial effects on these structurally inactive soils. However, plowing brings about only transient improvements and that the soil structure is easily degraded during the cropping phase. For initially compacted soils of East Africa Macartney *et al.* (1971) concluded that soil tillage could be limited to the row (seed zone) only. It is, therefore, concluded that some form of mechanical tillage is required for compacted and structurally inert soils of the semiarid region. This tillage may be performed once every other year, in the row zone only, or for every season depending on the antecedent soil conditions and crop requirements.

3. The Ridge–furrow System

The ridge–furrow system has been used and proved suitable for soil and water conservation in structurally unstable soils of the semiarid tropics. Ridge cropping is widely practiced with a range of variations to suit local needs. The variations range from ridges made along the slope to drain steep hills, across the slope to conserve soil and water, with cross ties in the furrows to hold surplus water, and with gentle gradients to facilitate water harvest. In the savanna regions of Nigeria and elsewhere in semiarid tropical Africa, crops are traditionally grown on ridges or small hillocks of mounds (Plate 7). This practice is useful on shallow soils to increase the effective root volume, on poorly drained soils to grow upland crops, on nutrient-deficient soil to heap up the fertile ash-rich topsoil, on steep slopes to provide a system for safe disposal of excess runoff, and in all conditions to facilitate harvest of roots and tuber crops.

The performance of the ridge–furrow system on soil and water conservation is variable. In Tanzania, Prentice (1946) reported higher yields from tie ridges than from flat seedbeds in years of partial drought. In contrast, Macartney *et al.* (1971) reported no beneficial effects of tie ridges on maize yields. Pereira *et al.* (1967) reported that tied ridges decreased risk of gully erosion more than the graded ridge–furrow system. Also in Tanzania, Dagg and Macartney (1968) observed that tie-ridged land produced significantly more maize yield only on red soil but not on a black (Vertisol) or an ash (Andosol) soil (Table 37). The ridge-furrow system reduced surface runoff and conserved more water in the root zone on the structurally unstable red soil only. With soils of high water holding capacity or permeability, the beneficial effects of a tie-ridge system are negligible. Similar conclusions were made by Honisch (1974) in Zambia. He reported that the highest grain yields of maize and sorghum were grown on a tie-ridge system. The results of experiments

Plate 7. Small hillocks on mounds are built by scraping the surface soil. Mounds facilitate harvesting of root and tubers.

conducted by IITA (1981) in Burkina Faso (formerly Upper Volta) also indicated that the highest yield was obtained in maize grown with a tie-ridge system (Table 38).

The effectiveness of ridge cropping also depends on duration, intensity, and time of occurrence of drought. In fact, it is a risk avoidance system and is effective on marginal lands and for regions prone to frequent drought stress.

4. Soil Guide to Tillage Requirements

Seedbed preparation and tillage requirements are soil specific and depend on soil constraints to crop production. No single tillage method can be recommended for all soils and crops. Responses to different tillage systems depend on soil properties, microclimate, rainfall characteristics, and the nature of the crops grown (Table 39). Tillage requirements for optimum crop growth vary according to the prevailing soil physical properties at the seeding time, and should be assessed in relation to soil strength, water status, soil temperature, and site characteristics. It is, therefore, relevant to develop a criterion to assess tillage requirements for seedbed preparation on the basis of antecedent soil properties and crop requirements.

Table 37. Maize grain yield (kg ha^{-1}) from a mechanized tie-ridge system for three soils in Tanzania

Soil	*Tie ridged*	*Ridge*	*Flat*	*Means*
Black soil	3274	3251	3085	3204
Red soil	3433	3029	2628	3030
Ash soil	1763	1113	1815	1563
Mean	2824	2465	2509	
LSD (.05)				
(i) Tillage	600			
(ii) Soils	347			

Reprinted with permission from Dagg and Macartney, *Exp. Agr.*, Vol. 4, 279–294. Copyright 1968, Cambridge University Press.

Lal (1983) developed a numerical rating system to assess a soil's suitability for the no-till system. This rating method is based on a few contrasting sites of experimentation and considers properties such as erosivity, erodibility, soil loss tolerance, rooting depth, available water-holding capacity, cation-exchange capacity, pH, and others. Some soils with good initial soil physical conditions require little or no tillage for satisfactory crop growth, i.e., biostructurally active soils. In other soils, with compacted surface layers and massive soil structure, mechanical tillage can bring about noticeable improvements in soil physical properties and resultant crop growth, e.g., as in biostructurally inactive

Table 38. Maize grain yield (kg ha^{-1}) in a tie-ridge system in Burkina Faso (IITA 1981)

	Management level		
Ridge system	*Low*	*High*	*Mean*
1. No earthing up	1040	1480	1260
2. Earthing up at 30 days after planting (DAP)	990	1470	1230
3. Earthing up at 30 DAP and tying the ridges every other furrow	1840	2540	2190
4. Earthing up at 30 DAP and tying all ridges	2040	3280	2660
Mean	1480	2190	1840
LSD (.05)			
(i) Management	701		
(ii) Ridges	416		

Table 39. Biophysical and social factors affecting crop response to different tillage methods

FACTORS

Soil	Micro- and mesoclimate	Crop	Socioeconomic
1. Texture and structure	1. Rainfall amount and distribution	1. Growth duration	1. Farm size
2. Surface characteristics	2. Length of dry season	2. Canopy characteristics	2. Degree of mechanization
3. Land form and slope	3. Erosivity	3. Root system	3. Availability of appropriate seeding and spraying equipment
4. Erodibility	4. Temperature regime and evaporative demand for the growing season	4. Consumptive water use	
5. Moisture-retention and -transmission properties	5. Radiation level	5. Soil-conserving vs soil-degrading crop	
6. Organic matter content		6. Biomass production	
7. Clay minerals		7. Drought susceptibility	
8. Effective rooting depth		8. Grain vs tuber crops	

soils. Whereas soils highly susceptible to erosion should be managed only by the no-till system, clayey soils with poor internal drainage respond only if the aeration status is improved. Soils with high silt and fine sand fractions are easily compacted and are prone to crusting, and those in the extremely arid regions may response favorably to chiseling in the row zone rather than plowing the entire field. On the basis of possible soil-related constraints, different tillage systems are suggested in Figures

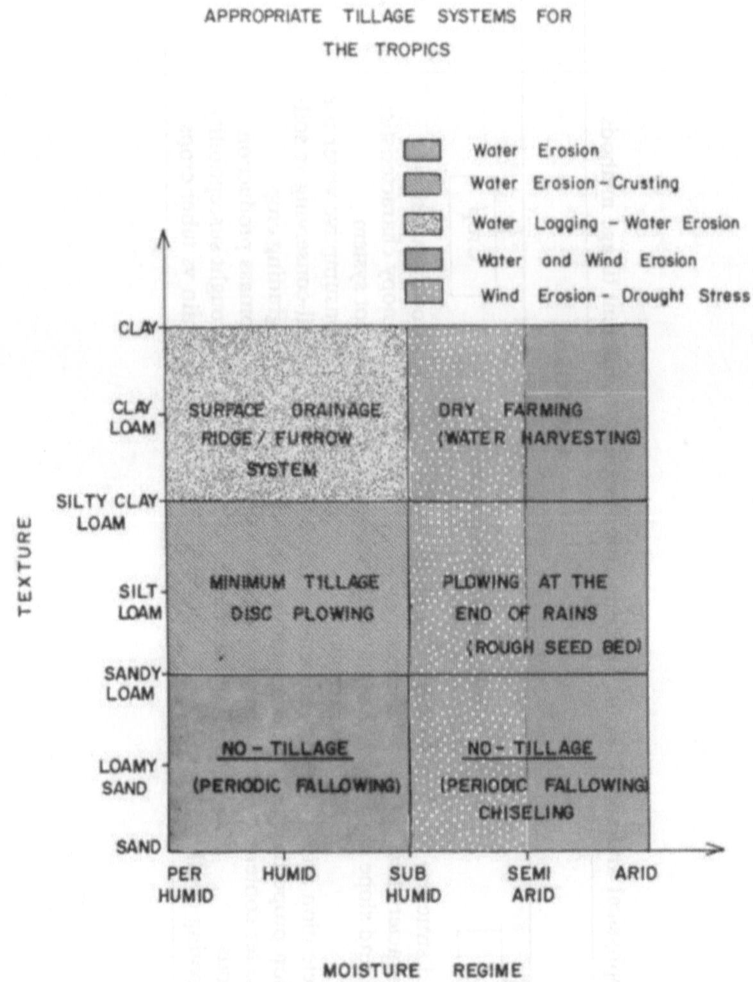

Figure 20. Appropriate tillage systems for the tropics in relation to soil texture and moisture regimes (Lal, 1983).

20 and 21. These generalizations are tentative guidelines subject to further validation and adaptation on different soils and ecologies. In the humid and subhumid tropics with soils of coarse texture in the surface horizon, no-tillage management can be applied with appropriate modifications to suit specific needs. In the semiarid regions some type of mechanical seedbed preparation is necessary. The frequency and the type of mechanical operations desired depends on soil characteristics and the crops to be grown.

C. Time of Planting

Timely planting is important to optimize the use of natural resources. The benefits of early planting have been widely documented in the

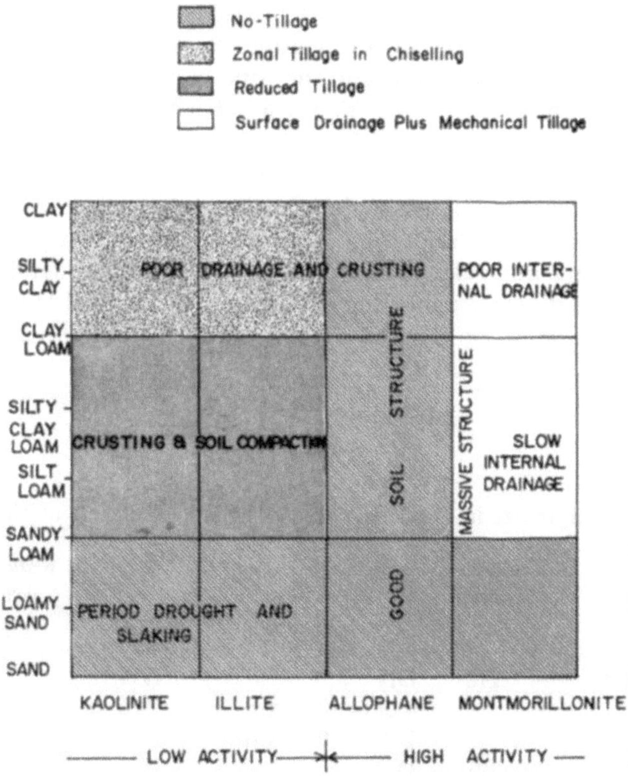

Figure 21. Tillage guide in relation to soil texture, clay minerals, and soil physical properties (Lal, 1983)

tropics and subtropics for a wide range of reasons, e.g., rainfall distribution, variety, soil moisture and soil temperature regimes, and mineralized nitrogen availability. The yield of a late-planted crop is always more variable and usually less than that of an early-planted crop (Lal, 1973). The effect of time of planting is especially critical in regions with a short rainy season. The yield of late-planted crop is adversely affected by the abrupt break in rains. For example, in Machakos, Kenya, Dowker (1971) related maize grain yield to the time of planting for three consecutive years from 1959 to 1962 (Table 40). The reduction in maize grain yield ranged from 4.7 to 6.3% for every day's delay in planting. The best yield was often obtained with seeding in a dry soil before the onset of the rains. Dry seeding prior to the onset of monsoons is a particularly beneficial technique for Vertisols with poor trafficability under wet conditions. Experiments conducted at ICRISAT, Hyderabad, India by Kampen *et al.* (1981) have shown that hitherto underutilized or unutilized Vertisols can be intensively cropped by "dry sowing"—seeding shortly ahead of the rainy season. This technique is particularly valuable for the regions where the rains commence fairly reliably.

In Nigeria Lal (1973) reported a significant interaction between the method of seedbed preparation and the time of planting for maize. In another study Lal (1975) found significant beneficial effects of early planting of yam on tuber yield (Figures 22 and 23). Yams planted in October yielded significantly more than those planted in February or April provided that the soil was mulched as a protection against high soil temperature and severe desiccation. The maximum soil temperature on unmulched heaps and ridges often exceeded 40°C at 5 cm depth, which adversely affected crop establishment (Figure 24). The soil temperature fluctuations are more on heaped up soil than on the flat soil surface (Figure 25).

Table 40. Yield of maize in Machakos, Kenya, as related to time of planting

Date	*Seasonal rainfall (mm)*	*Time of planting*	*Yield (kg ha^{-1})*	*Reduction in yield/day (%)*
Nov. 1959–1960	210	Dry	1979	—
		4 days after rain	1558	5.3
		7 days after rain	1170	5.9
Nov. 1960–1961	291	Dry	1526	—
		6 days after rain	954	6.3
April 1962	311	Dry	4580	—
		6 days after rain	3284	4.7

Adapted from Dowker, 1971.

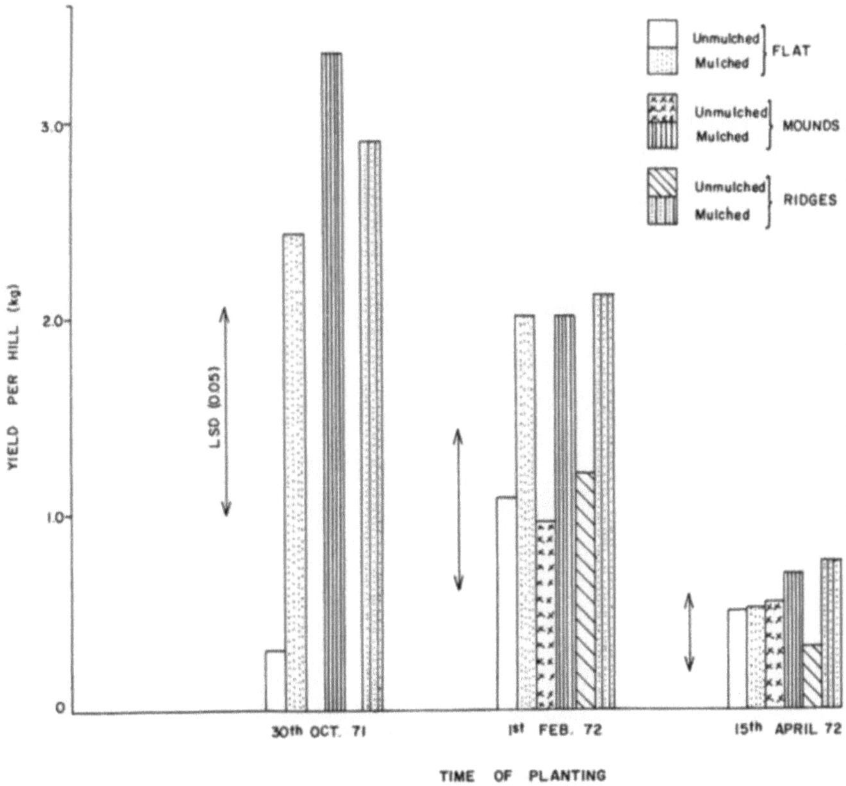

Figure 22. Effects of time of planting, methods of seedbed preparation, and mulching on yield of yam tuber per hill (Lal 1975)

D. Mulch Farming and Planted Fallows

Mulch farming is an important component in improving traditional agriculture and evolutionizing a permanently viable system of land use in the humid and subhumid tropics. Regular and substantial additions of crop residue mulch, left on the surface rather than incorporated into the soil, have proved to be beneficial for a wide range of soils and agroecological environments in the tropics. The main benefits include better soil and water conservation, improved soil moisture and temperature regimes, amelioration of soil structure, favorable soil turnover through enhanced biotic activity of soil fauna, and protection of the soil from high-intensity rains and from ultradesiccation. Because of soil amelioration and the effect of mulch on weed suppression, mulching is generally beneficial to crop growth and yields (Lal, 1975).

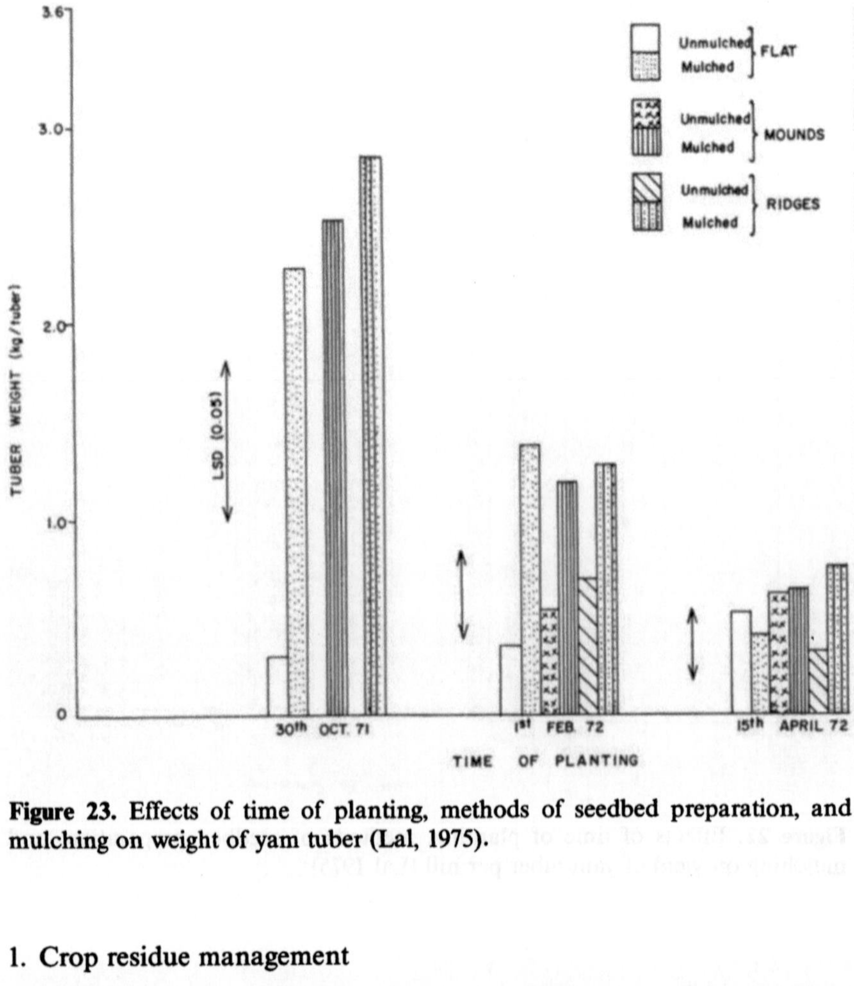

Figure 23. Effects of time of planting, methods of seedbed preparation, and mulching on weight of yam tuber (Lal, 1975).

1. Crop residue management

Mulching has a notable effect on the physical properties of tropical Alfisols. Lal *et al.* (1980) developed the following empirical relation between mulch rate (X in t ha^{-1}) and soil properties:

$$\text{Percentage water-stable aggregates } (>0.5 \text{ mm}) = 42 + 7.36X - 0.41X^2; \quad r = 0.98^{**}$$

$$\text{Dispersion ratio} = 26.9 \exp(-0.09X); \; r = 0.97^{**}$$

$$\text{Erosion ratio} = 71.9 \exp(-0.09X); \; r = 0.96^{**}$$

Lal and his colleagues reported significant improvements in infiltration rate as a consequence of improvements in structural stability by mulching. The infiltration rates were 32, 52, 60, 73, and 97 cm hr^{-1},

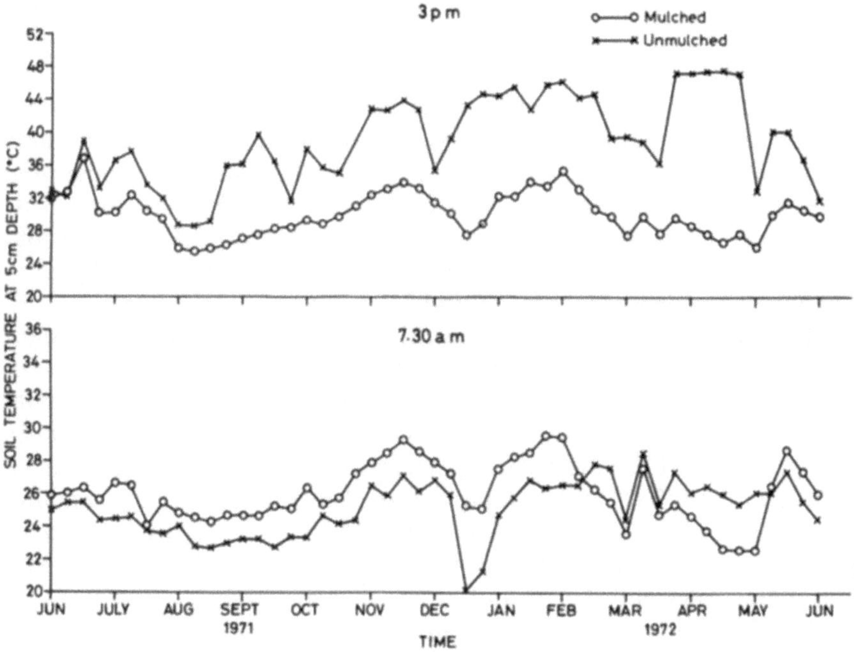

Figure 24. Seasonal fluctuations in soil temperature at 5-cm depth for mulched and unmulched Alfisol at IITA, Ibadan, Nigeria.

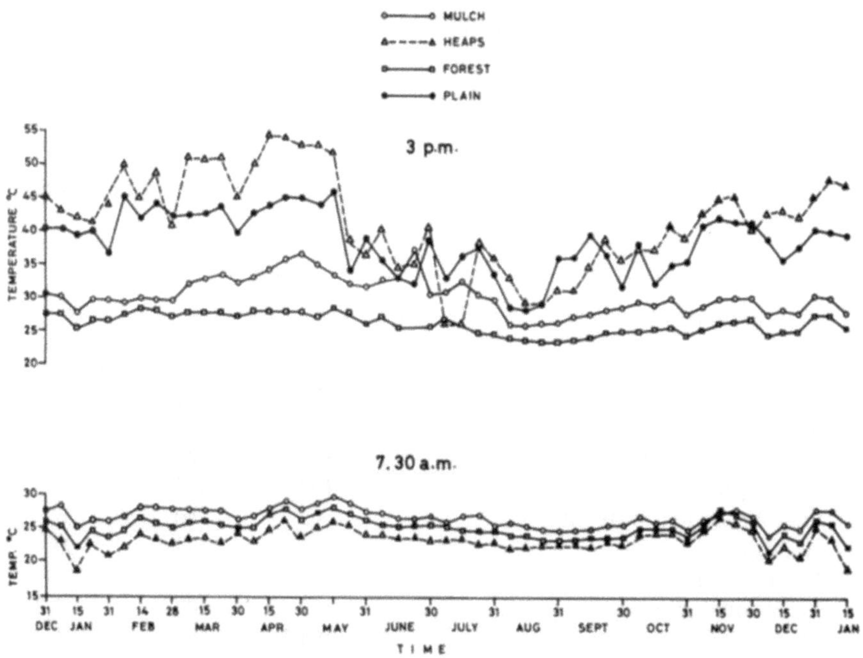

Figure 25. Soil temperature at 5-cm depth as influenced by methods of seedbed preparation and mulching Reprinted with permission from Lal, *Exp, Agr.,* Vol. 9 p. 310. Copyright 1973, Cambridge University Press.

respectively, for 0, 2, 4, 6 and 12 t ha^{-1} mulch rates. The high mulch rate also increased the available water-holding capacity. Soil moisture conservation is particularly improved through the use of plastic mulch (Maurya and Lal, 1981; Khatibu *et al.*, 1984). In addition to conserving soil moisture, mulching improves crop water-use efficiency through its regulatory effect on soil temperature and root growth. In the lowland tropics, straw mulch decreases the maximum soil temperature by 8° to 10°C at 5-cm depth (Figures 24 and 25).

The most beneficial effect of mulching is the reduction in water runoff and soil erosion. Losses of water by runoff decrease exponentially with increases in the mulch rate. Lal (1976) reported that the mean annual runoff losses from an Alfisol were 393, 81, 30, and 13 mm for mulch rates of 0, 2, 4, and 6 t ha^{-1}, respectively. The corresponding loss of soil was 11, 3.5, 0.5 and 0.3 t ha^{-1} for the slopes ranging between 1 and 15%. In another study Lal *et al.* (1980) observed that runoff losses were 18, 10, 4, 1, and 0% of the rainfall with corresponding soil losses of 10, 2, 0.5, 0.1, and 0 t ha^{-1} for mulch rates of 0, 2, 4, 6, and 12 t ha^{-1}, respectively.

Beneficial effects of mulch on runoff and soil erosion have also been reported for some soils in East Africa. Othieno (1975) and Othieno and Laycock (1977) reported notable reductions in runoff and erosion under young tea when crop residue mulch was applied (Table 41). The effectiveness of mulch in reducing runoff and erosion has also been demonstrated by research conducted in Zanzibar and Tanga, Tanzania. Khatibu *et al.* (1984) observed more water and soil loss from bare than

Table 41. Effects of mulching on runoff and erosion from a young tea plantation in Kenya

	Years after planting		
Variable/treatments	*1*	*2*	*3*
1. Ground cover (%)	1–30	30–60	60–70
2. Rainfall (mm)	2083	2045	1985
3. Runoff (mm)			
i. Manual weeding	181	127	32
ii. Herbicides	160	162	90
iii. Interplanted with oats	65	80	39
iv. Mulch	54	27	22
4. Erosion (t ha^{-1})			
i. Manual weeding	161	48	1.2
ii. Herbicides	168	81	6
iii. Interplanted with oats	35	4	0.4
iv. Mulch	0.5	0.1	0.08

Adapted from Othieno, 1975.

Table 42. Effects of mulching on runoff and soil erosion observed in Zanzibar

Observations	*Unmulched*	*Mulched*
Runoff (mm)	108.6	6.9
Runoff (% of rainfall)	10.2	0.01
Soil erosion (t ha^{-1})	5.5	0.2
Nutrient loss in runoff (kg ha^{-1})	28.0	2.0

Adapted from Khatibu et al, 1984.

from mulched plots (Table 42). In unmulched plots 10.2% of the total rainfall was lost as runoff compared to only 0.01% in the mulched treatment. The total soil loss from the mulched treatment was only 3.63% of that from the unmulched plots. At Tanga, Tanzania, Ngatunga *et al.* (1984) evaluated the effects of 6 t ha^{-1} of straw mulch on runoff and erosion in comparison with a bare, unmulched treatment. Their data in Table 43a show that mulching provided an effective erosion control even on steep slopes of up to 22%. The annual soil erosion from bare plots was 315, 488, and 490 times more and runoff 18, 23, and 29 times more than that from the mulched plots for 10, 19, and 22% slopes, respectively. In Taiwan Wang (1984) observed that soil erosion control from citrus orchards mulched with 10 t ha^{-1} of weeping love grass was as good as with level retention terraces (Table 43b). Residue mulch was even more effective than growing overall cover with bahia grass.

Research in Africa and elsewhere in the tropics has often shown more yield with mulching, particularly if the mulch material is a leguminous

Table 43a. Mulching effects on runoff and soil loss at Tanga, Tanzania

		Runoff (mm)		*Soil erosion (t ha^{-1})*	
Season	*Slope (%)*	*Bare fallow*	*Mulched*	*Bare fallow*	*Mulched*
Vuli	10	9.6	0.7	10.1	0.04
	19	10.8	1.2	26.1	0.06
	22	15.1	0.8	21.2	0.08
Masika	10	14.0	2.1	27.8	0.08
	19	48.9	1.7	66.7	0.11
	22	40.2	1.1	62.9	0.10
Annual total	10	23.6	2.8	37.9	0.12
	19	59.7	2.9	92.8	0.17
	22	55.3	1.9	88.1	0.18

Table 43b. Soil losses under different soil conservation treatments in young citrus orchards in Taiwan

Treatment	*Soil erosion*[a] *(t ha^{-1} yr^{-1})*
1. Clean cultivation	156.4
2. Level retention terrace	Traces
3. Mulching (10 t ha^{-1} with weeping lovegrass)	Traces
4. Strip cover (1.5-m wide) and mulching with bahia grass)	2.81
5. Overall cover and mulching with bahia grass	0.94

[a]Mean of 4 years.

Adapted from Wang, 1984.

residue. In Zaire, Jurion and Henry (1969) observed that residue mulching maintained soil fertility better than even the application of inorganic fertilizers (Table 44). The cotton (*Gossypium hirsutum*) yield on unmulched plots without fertilizer decreased to 12% of the initial yield after 10 consecutive cultivation years. In contrast, cotton yield on mulched plots without fertilizer was 87% of the initial yield and was about eight times more than that of the unmulched treatment. The beneficial effects of residue mulch on crop yields have also been demonstrated for Alfisols (Lal, 1975, 1978; Maurya and Lal, 1980) and Ultisols (Okigbo, 1965) in Nigeria (Okigbo and Lal (1980). The use of organic mulches as fertilizers is especially important for small landholders (1–2 ha) who neither can afford commercial fertilizers nor are sure of their effectiveness. Leguminous mulch materials can supply some nitrogen and other nutrients needed to boost the yield levels. Okigbo and Lal (1980) compared the yields of cassava, maize, cowpea, and soybean (*Glycine max*) under 22 different mulch materials. The data in Table 45 show that

Table 44. Effects of mulching and fertilizer application on relative cotton yield over a 10-year period in Zaire

	Unmulched		*Mulched*	
Year	*Without fertilizer*	*With fertilizer*	*Without fertilizer*	*With fertilizer*
1947–1948	(1039) 100	—	(1129) 100	—
1953–1954	19	(440) 100	99	(1434) 100
1955–1956	18	181	130	138
1956–1957	12	160	87	94

Figure in parentheses is the absolute yield in kg ha^{-1}.

Adapted from Jurion and Henry, 1969.

Table 45. Crop yield response[a] (t ha^{-1}) in 1976 under 22 different mulch treatments

Mulch	*Cassava (fresh roots)*	*Maize*	*Cowpea*	*Soybean*
Bare soil (control)	16.4def	3.0c	0.6a	0.6de
Maize stover	16.4def	3.3cd	1.1a	1.5abc
Maize cobs	17.8cdef	3.3cd	1.1a	1.4abcd
Oil palm leaves	17.1def	3.2cd	1.2a	0.9bcde
Rice straw	17.9cdef	3.5bcd	1.0a	1.5abc
Rice husks	28.3a	3.7abc	1.1a	0.8de
Kikuyu grass straw (*Pennisetum*)	14.2ef	3.3cd	1.2a	1.4abcd
Elephant/napier grass	16.6def	3.3cd	0.9a	1.3bcd
Guinea grass	15.5f	3.6bcd	2.1b	1.5ab
Andropogon straw	18.5cdef	3.5bcd	1.0a	1.2bcde
Cattail straw (*Typha*)	16.7def	3.1cd	1.0a	1.1bcde
Cassava stem (chipped)	20.9cd	3.8abc	0.9a	1.4abcd
Pigeon pea tops	22.9bc	3.7abc	1.1a	0.9cde
Pigeon pea stem (chipped)	19.9cde	3.5bcd	1.0a	1.3bcd
Legume husk	26.4ab	4.4a	1.0a	1.5abc
Soybean tops	22.9bc	4.2ab	1.0a	1.2bcde
Hemp (*Eupatroium*)	18.8cdef	3.6abc	1.0a	1.2bcde
Mixed twigs (chipped)	18.5cdef	3.4bcd	1.0a	1.2bcde
Sawdust	20.5cde	3.7abc	0.9a	1.9a
Black plastic	30.5ab	3.0cd	0.9a	1.1bcde
Transluscent plastic	27.7ab	2.7d	1.0a	1.1bcde
Fine gravel	22.9bc	3.1cd	1.0a	1.0bcde

[a]Figures followed by similar letters are statistically similar within vertical rows.
Adapted from Okigko and Lal, 1980.

rice husk increased maize yields by 0.7 t ha^{-1} and cassava by 12 t ha^{-1}. On an Ultisol in southeast Nigeria Maduakor *et al.* (1984) observed that mulching significantly increased the yam tuber yield (Table 46a). The yam tuber yield increased by 20% on both ridge (from 12.8 to 15.4 t ha^{-1}) and flat (from 13.4 to 16.1 t ha^{-1}) planting. The diameter and length ot yam tubers were also increased by residue mulching. On an Ultisol in southeast Nigeria Hulugalle and Lal (1984, unpublished data) observed that the yield of cocoyam (*Xanothomonas* spp.) tubers was increased from 2.8 t ha^{-1} to 5.75 t ha^{-1} by mulching. It is well known that organic farming and the use of crop residue mulches have been the basis of sustained intensive, agriculture in China and other tropical countries of Southeast Asia.

Table 46a. Effects of tillage methods and mulching on yield and yield components of yam tubers

Treatment	*Diameter (cm)*	*Length (cm)*	*Number*	*Mean tuber yield (t ha^{-1})*
Ridge, mulch	16.7 a	23.5 b	9,936 b	15.4 a
Flat, mulch	17.1 a	21.7 a	12,916 a	16.1 a
Ridge, no mulch	13.3 b	20.7 a	10,385 b	12.8 b
Flat, no mulch	13.8 b	20.6 a	10,128 b	13.4 b

[a]Figures followed by similar letters are statistically similar within vertical rows.
Adapted from Maduakor et al, 1984.

The effects of organic mulches have not been widely explored in tropical America. Sanchez and Salinas (1981) observed only slight increases in the yield of maize mulched with *Melinis mintriflora* compared with an unmulched control on a Typic Haplustox near Brasilia, Brazil. In another study at Yurimaguas, Peru, Sanchez and Salinas reported positive effects of mulching with *Panicum maximum* on grain yield of many crops grown on Typic Paleudults. Mulching increased the yield of maize by 10.7%, of soybeans by 2.2%, and of peanuts by 2.8%. The yields

Table 46b. Yield response (kg ha^{-1}) of maize and cowpea to different mulch materials and mineral fertilizer in an Eastern Amazon Oxisol. CAPITAO POCO (CPATU) PARA, BRAZIL, 1983

Mulches used (10 t ha^{-1}) of d.m.)	*First crop (maize)*[a]		*Second crop (cowpea)*[b]	
	NPK (kg/ha) 120-80-60	*NPK 0-0-0*	*NPK 30-80-60*	*NPK 0-0-0*
Elephant grass	4646	2144	1227	80
Pueraria	5697	3342	1187	114
Weeds	4911	2215	1394	105
Sec. forest (2 to 3 years)	4462	1560	1191	35
Sec. forest (4 to 5 years)	4479	1807	1397	95
Rice husks	4398	1146	1487	163
Maize cobs + husks	4863	2101	1302	41
Bare soil	3539	78	1169	7
LSD. 5%	987		281	
LSD. 1%	1321		376	
LSD. 0.1%	1729		493	

[a]Grain moisture content 14.5%.
[b]Grain moisture content 13%.
Adapted from Schöningh and Alkämper, 1984.

Table 47. Effects of legume and grass covers on physical properties of a tropical Alfisol

Cover crop	*Bulk density* $(g\ cm^{-3})$	*Infiltration*[a] *capacity* $(cm\ 3\ hr^{-1})$	*Infiltration*[a] *rate* $(cm\ hr^{-1})$	*Organic carbon (%)*		*CEC (meq per 100 g)*		*Equilibrium soil moisture content (%)*		
				B[b]	*A*[c]	*B*[b]	*A*[c]	*0 to 5 cm*	*0 to 10 cm*	*11 to 15 cm*
Brachiaria	1.34	65	19	1.21	1.57	6.1	8.5	10.1	10.6	10.3
Paspalum	1.35	65	14	1.23	1.45	7.1	8.2	9.7	11.4	10.3
Cynodon	1.30	66	18	1.30	1.70	6.9	8.9	14.8	11.5	8.9
Pueraria	1.32	76	16	1.27	1.50	6.6	7.7	20.1	12.1	12.1
Stylosanthes	1.33	74	16	1.30	1.63	6.3	8.8	18.5	15.9	15.6
Stizolobium	1.33	77	21	1.30	1.57	6.7	10.5	14.7	14.2	11.7
Psophocarpus	1.14	124	42	1.20	1.57	5.9	10.9	21.2	15.8	13.6
Centrosema	1.33	72	18	1.30	1.53	6.5	10.0	15.9	15.0	15.5
Control	1.42	71	13	1.33	1.37	8.2	8.4	11.0	12.7	12.0
LSD (.05)	0.04	66	9	0.50	0.23	2.3	3.5	6.2	2.3	3.1

[a]Preceeding infiltration capacity was 44 cm per 3 hr and infiltration rate was 17 cm hr^{-1}.

[b]B = in 1974 before seeding cover crops.

[c]A = 1976 2 years after.

Adapted from Lal *et al*, 1979.

of upland rice and cowpea, however, were slightly suppressed. These authors reported that mulching with *Pueraria phaseoloides* straw was more beneficial than with *Panicum* and produced 80% of the maximum yields without the addition of chemical fertilizers. Schöningh and Alkämper (1984) studied yield response of maize and cowpea to different mulch materials in an Eastern Amazon Oxisol in Brazil. Their data (Table 46b) show that mulches with high decomposition rates (low C:N ratio) affected the yields of maize more than those resistant to decomposition (high C:N ratio). The reverse was the trend, however, with cowpea. In addition to yield, mulches positively influenced soil mesofauna, organic matter, and CEC. The soil moisture reserves in the 0- to 10-cm layer was generally more on mulched than unmulched plots, and mulching drastically decreased the soil surface temperature.

2. Cover Crops and Planted Fallows

When crop residue mulch is inadequate, a practical means to procure mulch is by the incorporation of an appropriate cover crop or planted fallow in the rotation. In addition to providing residue mulch, planted fallows are more effective in restoring soil physical and nutritional properties than long bush fallows (Tables 47 and 48). Organic matter can be built up and the soil structure improved even on eroded and degraded lands by growing appropriate planted fallows for 2 to 3 years (Wilkinson, 1975; Juo and Lal, 1977; Lal *et al.*, 1978, 1979; and Wilson *et al.*, 1982). For example, the data in Figure 26 show that compared with the preseeding control, water infiltration of an eroded Alfisol was significantly improved by 2 years' growth of *Psophocarpus*, *Centrosema*, *Pueraria*, *Stizolobium*, and *Stylosanthes*. The mean infiltration rate increased in 2 years of fallowing from 9 cm hr^{-1} to 17 cm hr^{-1} under grasses and 23 cm hr^{-1} under legumes. Also increased were the soil organic matter content, available phosphorus, and exchangeable cations (Table 49). Growth is generally better after a planted fallow of legumes than after grass fallows (Lal *et al.*, 1978). Recent studies at IITA by Wilson and others (IITA, 1983) have shown the beneficial effects of in situ mulch from the cover crop *Mucuna utilis*. Maize grown without *Mucuna* mulch was stunted and prone to drought, suffered from high soil temperatures, and produced less grain even with the application of 60 and 90 kg N per hectare.

Leguminous crops have long been grown in association with plantation crops. The most commonly used legumes for this purpose are *Calapogonium muconoides*, *Centrosema pubescens*, *Indigofera spictata*, *Pueraria phaseoloides*, *Stylosanthes guianensis*, *Dolichos hosel*, and *Glycine javanica*. The beneficial effects of grass and leguminous covers are also well established for soils of the semiarid regions of West Africa (Pereira *et al.*, 1954). Pereira *et al.* (1958a,b) confirmed that fallowing with teff grass

Table 48. Effects of cover crops on soil moisture characteristics of an Alfisol in Western Nigeria

	Soil moisture retention (%) at different suctions (bars)															
	0		*0.1*		*0.3*		*0.5*		*1*		*2*		*3*		*15*	
Cover crops	*B*[a]	*A*[b]	*B*[a]	*A*[b]	*B*[a]	*A*[b]	*B*[a]	*A*[b]	*B*[a]	*A*[b]	*B*[a]	*A*[b]	*B*[a]	*A*[b]	*B*[a]	*A*[b]
Bracharia	32.0	37.5	17.5	18.7	13.0	12.6	11.8	11.8	11.1	11.3	10.5	9.4	9.0	9.1	7.7	8.3
Paspalum	35.7	40.9	17.5	19.5	14.2	14.3	12.4	12.7	11.8	11.1	10.7	11.2	9.9	10.0	8.7	9.3
Cynodon	36.1	41.3	18.3	19.7	14.3	13.5	13.6	12.5	12.4	10.5	10.3	9.3	9.9	8.5	8.8	9.9
Pueraria	31.7	39.7	16.4	18.4	14.4	14.0	12.7	12.9	12.0	12.1	10.9	10.8	9.8	9.2	8.9	9.2
Stylosanthes	31.5	40.3	18.3	19.1	14.6	14.6	13.6	13.0	12.9	11.1	10.9	10.2	9.9	9.2	9.4	9.5
Stizolobium	32.9	40.2	17.4	19.0	15.1	14.2	13.8	13.7	12.8	11.5	11.7	10.6	10.7	9.6	8.8	9.7
Psophocarpus	34.2	38.3	16.2	18.8	14.8	14.2	13.1	13.7	12.6	11.2	11.6	11.4	10.3	10.0	9.2	9.7
Centrosema	34.1	38.1	17.3	18.4	15.1	15.9	13.9	14.2	13.0	11.3	11.9	11.2	10.4	9.9	9.0	9.9
Control	30.6	37.6	15.3	15.7	15.2	14.5	13.9	12.9	13.1	11.4	11.7	10.9	10.3	8.9	8.6	8.9
LSD (.05)	4.7	3.9	9.7	2.7	3.0	2.8	2.5	2.6	2.7	2.0	1.9	1.8	2.2	2.1	2.5	1.9

[a]B = 1974 before seeding cover crops.

[b]A = 1976, 2 years after.

Adapted from Lal *et al,* 1979.

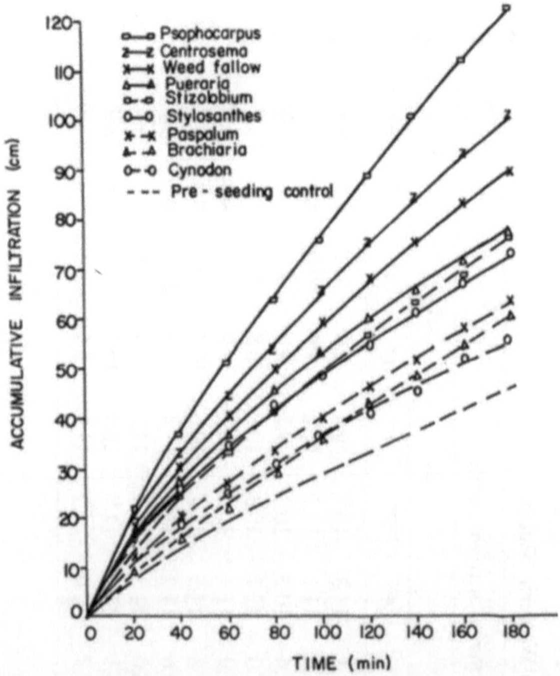

Figure 26. Effects of 2 years of grass and leguminous fallow on the infiltration characteristics of a severely eroded tropical Alfisol (Lal *et al.*, 1978.)

Table 49. General properties of the surface layer (0 to 15 cm) of cultivated degraded Alfisol in comparison with that under forest cover

Properties	*Under forest cover*	*After 5 years of cultivation*	*After 2 years of cover crops*	
			Grasses	*Legumes*
pH (H_2O)	6.5	5.5	5.2	5.6
Clay (%, w/w)	18.4	21.2	—	—
Silt (%, w/w)	10.3	9.5	—	—
Sand (%, w/w)	71.3	69.3	—	—
Gravel (%, w/w)	5.0	20.0	—	—
Organic C (%, w/w)	1.7	1.2	1.5	1.6
Total N (%, w/w)	0.16	0.12	0.18	0.20
Bray-P (g/g)	15	37	60	33
Exchangeable bases (meq per 100 g)				
Ca^{2+}	4.8	4.0	6.2	6.7
Mg^{2+}	1.3	0.80	1.3	1.6
K^+	0.6	0.4	0.9	0.9
Bulk density (g cm^{-3})	1.00	1.50	1.33	1.29
Infiltration rate (cm h^{-1})	100	9	17	23

Adapted from Wilson *et al*, 1982.

Table 50. Effects of cropping systems on free-draining pore space of surface soil

Cropping Systems	*Pore space (% by volume) at different tensions (cm of water)*			*Soil temperature at 25-cm depth (°C)*
	10	*50*	*100*	
Bare fallow	7.4	22.0	25.7	50
Crops and grasses	11.5	26.1	28.1	23–35
LSD (.05)	4.1	5.2	4.9	

Reprinted with permission from Pereira et al., *Emp. J. Exp. Agr.*, Vol. 26, 203–228. Copyright 1958, Cambridge University Press.

(*Eragrostis abyssinica*) improved rainfall infiltration and decreased runoff and erosion (Table 50). Planted fallows are increasingly being turned to as an integral component of rotations based on seasonal food crops. Cover crops that have been found useful as planted fallows before the no-till production of grain or food crops are: *Centrosema pubescens, Pueraria phaseoloides, Stylosanthes guianensis, Psophocarpus palustris,* and *Mucuna utilis*. The effects of various legume covers on the following grain and root crops at IITA are shown in Table 51. If grass covers can be effectively suppressed, they can also improve crop yields over the weed-fallow control, e.g., *Setaria, Brachiaria, Melinis,* etc.

A fallow management system of growing grain/food crops through the unkilled mulch on the principles of mixed cropping is termed "live mulch" (Lal, 1975). A small strip is opened, with or without using herbicides, to grow a seasonal food crop. The system works if the fallow crop is not competitive for light, moisture, and nutrients. Wilson *et al.* (1982) observed severe yield reductions by unsuppressed climbing covers. With a low-growing nonclimber, favorable grain yields can be obtained with suitable agronomic management (Akobundu 1980, 1982) (Table 52). In regions with rainfall of less than 2000 mm year^{-1} and with periodic drought stress in the growing seasons, live mulch is likely to compete for soil moisture and suppress the yield of food crops grown through it. Observations on soil moisture under different live mulches made at IITA show that some deep-rooted live mulches can deplete soil water from up to 120-cm depth (Figure 27). The data in Table 53 comparing soil moisture content for 0 to 5 cm depth on an Alfisol at IITA under different live covers with that under residue mulch and bare fallow treatments show highly competitive nature of most live covers for limited soil moisture reserves. Unless properly killed by chemical or mechanical means, live mulches may become the most aggressive weeds and completely devour the grain/food being grown through them.

Table 51. Grain yield (t ha^{-1}) as affected by various sod species, 1974

	Cropping sequence			
	Maize–cowpeas	*Cowpeas–maize*	*Pigeon peas–soybeans*	
A. First season	Maize	Cowpea	Pigeon pea	
Panicum	3.13	0.37	1.19	
Setaria	5.77	0.49	0.86	
Brachiaria	5.17	0.76	1.27	
Melinis	5.18	0.63	1.19	
Centrosema	5.79	0.76	1.23	
Pueraria	4.77	0.84	1.12	
Glycine	5.05	0.65	1.04	
Stylosanthes	5.17	0.61	1.17	
Plowed	4.87	0.50	1.06	
LSD (.05)	2.21	0.05	0.20	
B. Second season	Cowpea	Maize	Soybean	Cassava
Panicum	0.62	1.69	0.50	3.50
Setaria	0.71	2.97	0.91	7.90
Brachiaria	1.04	3.80	1.14	17.39
Melinis	0.87	3.43	0.77	18.05
Centrosema	0.76	3.73	0.75	15.01
Pueraria	0.79	3.44	0.80	19.49
Glycine	0.71	3.02	0.93	14.12
Stylosanthes	0.67	3.11	0.91	19.83
Plowed	0.43	2.06	0.51	8.05
LSD (.05)	0.06	0.53	0.23	2.53

Adapted from Lal *et al*, 1978.

Table 52. Grain yield of maize (t ha^{-1}) as affected by ground cover in a live mulch crop production system

Ground cover	*Grain yield*
No tillage	2.31
Conventional tillage	2.54
Arachis	2.78
Centrosema	2.67
Psophocarpus	2.59
Least significant difference (= 5%) between means:	
Within type of ground cover	0.91
Between different types of cover	0.90

Adapted from Akobundu, 1980.

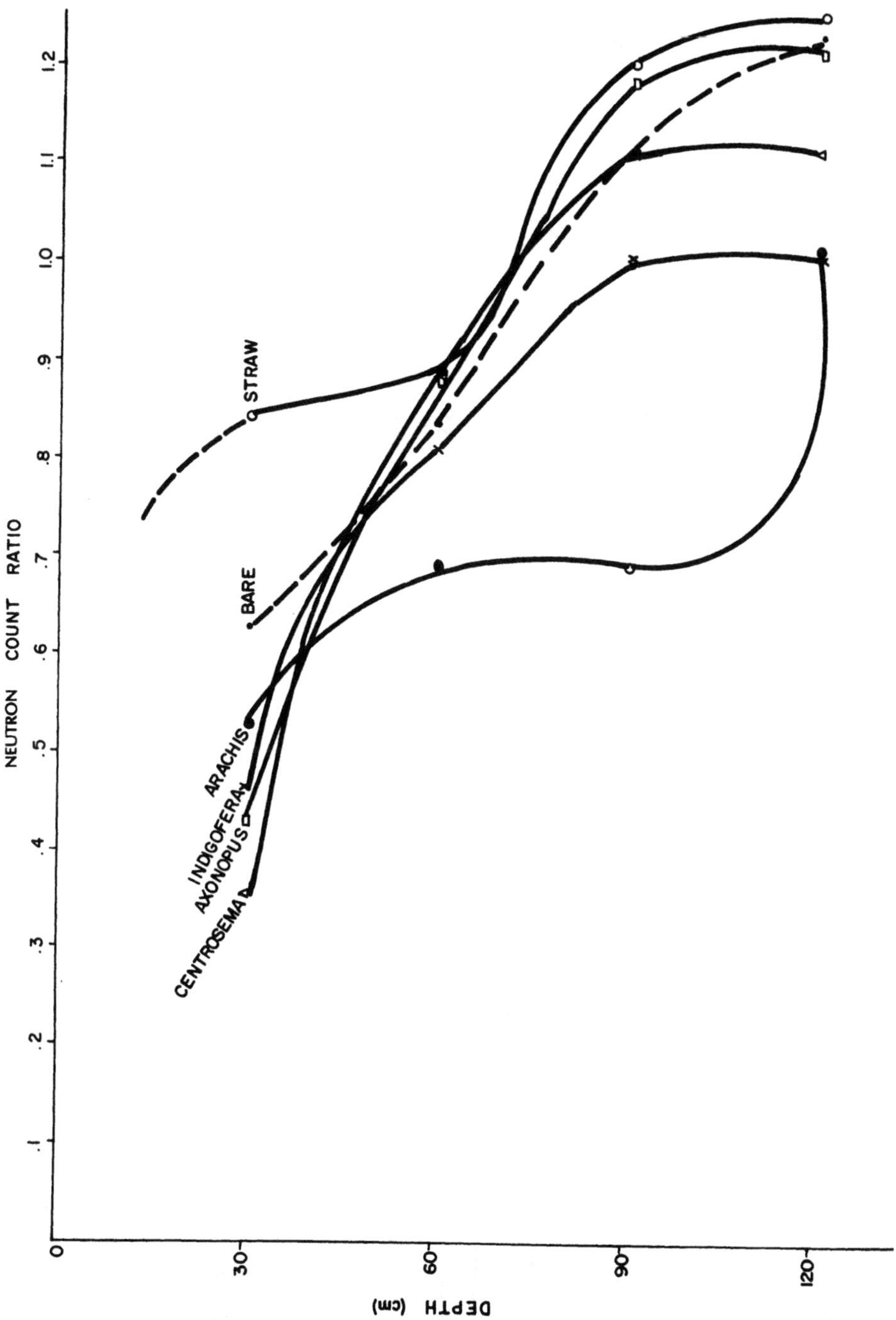

Figure 27. Soil moisture reserves under different crop covers, bare soil, and crop residue mulch.

Table 53. Gravimetric soil moisture (%) measurements (0 to 5-cm depth) under different cover crops and surface mulches during the dry season

Cover crop	*11/11/76*	*18/11/76*	*24/11/76*	*30/11/76*	*6/12/76*	*14/12/76*
(a) 0 to 5 cm						
Straw mulch	17.0	12.4	7.6	7.5	8.1	7.05
Bare	16.0	10.9	7.2	6.5	6.7	6.9
Desmodium	9.8	7.5	3.4	3.4	2.8	2.0
Arachis	16.6	7.9	5.4	5.3	4.3	3.5
Indigofera	16.5	10.8	5.3	3.9	5.5	5.3
Psophocarpus	13.2	9.4	6.9	5.8	5.3[a]	5.5[b]
Paspalum	12.8	8.7	6.9	2.9	2.7	2.4
Centrosema	12.9	8.8	6.3	6.5	6.2	5.5
Axonopus	15.1	7.8	3.0	3.5	3.9	3.9
(b) 5 to 10 cm						
Straw mulch	14.0	12.2	8.2	7.3	8.7	6.0
Bare	12.4	10.5	7.9	7.5	7.5	5.4
Desmodium	13.0	8.4	4.0	4.7	2.4	1.9
Arachis	14.8	8.4	5.2	5.4	3.5	3.2
Indigofera	10.2	6.1	3.9	5.7	6.1	4.9
Psophocarpus	11.3	10.3	4.7	8.0	4.2	5.4
Paspalum	12.0	8.3	5.0	4.0	4.6	2.7
Centrosema	10.2	8.1	7.2	6.5	5.8	5.4
Axonopus	11.7	6.2	3.6	3.2	3.8	3.9

[a]Mowed.

E. Cropping Systems and Agroforestry

Traditional farming systems are stable because of their diversity—many crop species of varying growth cycles, canopy structures, root systems, and water and nutrient requirements are grown simultaneously. Different forms of intercropping traditionally used include growing annual food crops in association with tree crops, plantation crops, and pastures, or with other annual crops. The spatial and temporal arrangements are designed to optimize the use of natural resources and insure continuous food flow (Okigbo, 1978; Willey, 1979).

Mixed cropping, as it is practiced in the traditional farming system, should feature strongly in any attempt to evolutionize the system. The benefits of mixed cropping are generally greater under adverse conditions with low inputs than in regions of few constraints and high inputs. It is a risk-avoidance practice. The benefits of mixed cropping in decreasing erosion losses are well established (Aina *et al.,* 1977). In a recent study on an Alfisol at IITA, Ibadan, Nigeria, Hulugalle and Lal (unpublished data, 1984) observed that the water-use efficiency of a maize + cowpea intercrop system was about twice as much as that of monocropped maize or cowpea. The water-use efficiency for the maize and cowpea grown in the same row was greater than that in alternate rows (Table 54). Under conditions of limited water supply, an appropriate crop mixture insures some yield rather than a complete failure.

Agroforestry is a special case of mixed cropping, which involves growing of a deep-rooted perennial leguminous shrubs and trees, and minimizes the soil degradation risks of an intensive arable land use (Pereira *et al.,* 1954; King, 1979; Vergara, 1982; Okigbo and Lal, 1980). Perennial crops that fruit annually, such as banana, are more suitable for intercropping with annuals than plantation crops. Shade-tolerant staples, such as cocoyam, are grown in association with plantation crops, e.g.,

Table 54. Energetics and water use efficiency of maize–cowpea intercropping systems

Cropping systems	Water-use efficiency (kg/mm/ha)	Energy ($GJ\ ha^{-1}$) Input	Output	Output/input
Maize monocrop	2.1c	2.0	19.5	9.8
Cowpea monocrop	2.1c	1.3	19.0	14.6
Maize + cowpea (same row)	4.2a	1.7	38.4	22.6
Maize + cowpea (alternate rows)	3.6b	1.7	32.3	19.0

Unpublished data of Hulugalle and Lal, 1984.

cocoa, especially in the early stages of plantation establishment or along the outer margins. Less shade-tolerant cassava is commonly grown in association with open-canopy coconut.

Experiments conducted at IITA, Ibadan, Nigeria, have demonstrated the usefulness of alley cropping—the practice of growing food crops between rows of specially planted woody shrubs or trees. The woody shrubs are regularly pruned during the cropping season to prevent shading, provide mulch, and reduce water use by the perennials. Either an alternate row or alternate strip system of tree arrangements can be used. The trees are pruned at the appropriate height to promote rapid regeneration (Kang *et al.*, 1981). The practice is particularly useful for smallholders because the legume prunings are estimated to supply an equivalent of 30 to 40 kg ha^{-1} of N to the grain crops grown between the alleys. Vergara (1982) described a desirable perennial species as having rapid growth, nitrogen-fixing capability, root system and a compatible with food crops. Legume species commonly grown in association with food crops are *Sesbania, Glyricidia, Leucaena, Parkia, Ramon, Calliandra, Albizia, Acacia, Cassia, Pithecellobium, Mimosa, Prosopis,* and *Samanea*. The choice of an appropriate species depends on soil, climate, and seasonal crops to be grown.

Using the prunings of the perennial shrub as a mulch for no-till food crop production between the rows of shrub is an orderly improvement of the traditional farming system where food crops are grown scattered in between the standing trees. If the rows of perennial shrubs are grown on contour, the hedges so formed can reduce risks of soil erosion and gully development. Deep-rooted perennials also improve water infiltration into the soil by improving the macroporosity of the subsoil horizon (Juo and Lal, 1977). In the Philippines, for example, Baconguis and Rondilla (1979) observed that the infiltration rate of soil planted to *Leucaena leucocephala* was 4.25 L min^{-1} compared with 1.98 L min^{-1} in regions where seasonal crops were grown. Improved infiltration rates, reduced runoff volume and velocity, and prevention of soil splash by mulch prunings reduce soil erosion on land managed with contour-planted hedgerows of woody shrubs and perennials. The data on runoff and soil erosion from field runoff plots (70 m × 10 m) established on a 7% slope on an Alfisol at IITA are shown in Table 55. Three-year-old *Leucaena* hedges planted at 2-m intervals reduced runoff and erosion equivalent to that of the no-till mulch plot. Although the maize grain yield was only slightly less, the cowpea yield with *Leucaena* alleys was about 50% of the no-till control. About 15% of the land goes out of production under hedgerows at spacings of 4 m and on slopes of about 10%. For the alley system to have yields equivalent to the no-till mulch system, it must either yield 15% more per unit area or contribute an equivalent amount in nitrogen or fuel over and above the extra labor involved. There is

Table 55. Effect of alley spacing and species on runoff, erosion, and yield

Treatment	*Runoff (mm)*	*Erosion (t ha⁻¹)*	*Grain yield (t ha⁻¹)*	
			Maize	*Cowpea*
Plowed	232	14.9	3.6	0.58
No-till	6	0.03	4.0	1.19
Glyricidia (4 m)	38	1.7	3.6	0.67
Glyricidia (2 m)	20	3.3	3.3	0.50
Leucaena (4 m)	10	0.2	3.7	0.58
Leucaena (2 m)	13	0.1	3.8	0.45

considerable scope for improving the agronomic management of the systems.

The usefulness of various inter- and mixed-cropping systems for acid Oxisols and Ultisols of the Amazon region has been discussed in detail by Sanchez and Salinas (1981). In spite of the obvious ecological benefits of these practices these authors concluded that these systems would remain unproductive unless fertilizers and lime were added to at least partially overcome critical acid soil constraints. The acid soils of the Amazon Basin are critically deficient in P and Ca, and it appears that the soil does not respond to other inputs unless these elements are supplied in adequate amounts.

F. Mixed Farming

Integrating livestock with seasonal and perennial crops is a useful approach to providing diversity and effectively utilizing the limited resource. Mixed farming is also an appropriate means to utilize planted fallows economically and provide compost to replenish soil fertility. Lightly grazed pasture and browse legumes incorporated in rotation with nitrophilous grain crops may be an effective strategy to supply nitrogen to the crop. In tropical Australia McCown *et al.* (1979) have developed a system involving enhanced production of legume fodder in the cropping enterprise. Integrating livestock with crop production is a means to enhance the economic use of planted fallow. McCown *et al.* (1984) evaluated the nitrogen contribution of a pasture-based system using maize or sorghum as a bioassay. Their data in Figure 28 show that the maize grain yield with no N fertilizer following 1-year leys of various pasture legumes was equivalent to that on plots receiving at least 50 kg N per hectare following 1 year of grass. These authors observed that legume leys of longer duration had greater effects on the first crop and a greater residual effect on the second crop. Care must be taken, however, not to

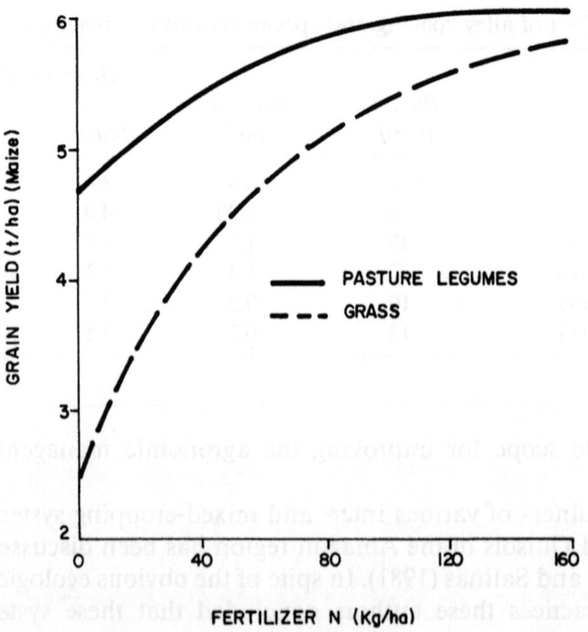

Figure 28. Nitrogen contribution of a pasture-based system to maize grain yield in Northern Australia (McCown *et al.*, 1984).

overgraze the pastures. Excessive grazing not only diminishes the nitrogen contribution but also can cause soil compaction, thereby reducing the possibility of growing the following grain crops by the no-till system. A low stocking rate or the cut and feed system are obviously desirable practices.

VIII. An Agroecological Approach to Improving Traditional Agriculture in the Tropics

High-input agricultural technology is useful when the basic ingredients are locally available, affordable, and maintainable. For subsistence farming on small landholdings a stepwise improvement based on low inputs is not only technically feasible but also culturally, economically, and ecologically desirable. Improved and efficient farming systems must, therefore, sustain high levels of production with minimal requirements for chemical fertilizers and other fossil fuel-based inputs. Kiley-Worthington (1981) defined an ecological agricultural system as "self-sustaining low input, economically viable, small farming systems managed to maximise production without causing large or long term changes to the environment or being ethically or aesthetically unacceptable." This concept is currently applicable to the subsistence farmer of the tropics and subtropics, and the various subsystems or components to develop this strategy have been outlined in the previous section.

The first and the easiest step to improve the traditional system is to introduce high-yielding cultivars and incorporate them into compatible and organized cropping systems including integration with woody perennials. The next step is to improve land clearing and tillage methods to facilitate mechanization, alleviate labor constraints, and improve soil and water conservation. Soil fertility maintenance is built in through frequent use of legume plants in rotation and growing annuals with woody shrubs as a continuous source of nitrogen-rich mulch.

One of the objectives of this approach is to economize energy use. This is where impressive progress can be made in developing and using appropriate technology, e.g., simple but efficient tools, either manually powered or animal driven, for land clearing, tillage, seeding, fertilizer application, and harvesting. Tools that can use renewable sources of energy and draft animals are preferred. The so-called traditional systems based on human muscle as the chief source of power are not necessarily efficient if the improved devices are not used to achieve the task more efficiently.

The energetics of all inputs and outputs for six different systems of land development and soil management studied at IITA, Ibadan, Nigeria, are shown in Table 56 for a 3-year rotation. The first crop after land clearing (1979) did not receive any fertilizer. The traditional system, based entirely on human energy (slash, burn, plant, weed, and harvest), was the least productive. The mechanized systems, most destructive in terms of accelerated soil erosion and compaction, were also the most efficient. For the cassava crop harvested in 1980, however, the efficiency of the system involving manual clearing followed by no-till planting increased drastically. The maximum output:input ratio for 1980 was attributed to a high cassava tuber yield in spite of low inputs. The cassava tuber yield of 15 to 16 t ha^{-1} was obtained with little fertilizer input. By 1981, the traditional system of planting, though less productive, was more efficient than some of the completely mechanized systems. A drastic decline in the output:input ratio of the tree pusher/root rake clearing followed by conventional tillage was due to yield reduction caused by severe soil erosion and the erosion-induced degradation in soil quality.

It is important to note that the output:input ratio decreased with time in most systems that involved mechanical land clearing and the conventional tillage methods. Improvements should be sought in the appropriate tools and cropping systems so that not only is the production increased with low inputs but it is also sustained. It is, however, only through the application of the sound basic knowledge of soil, climate, agronomy and social conditions of the region that up to date and modern skills of engineering and resource management can be used to develop appropriate technology to evolutionize farming systems gradually. It is, however, a continuously evolving system. Once the production increases,

Table 56. Energetics of land-clearing and tillage methods for a 3-year maize–cassava–maize–cowpea rotation investigated on a 3- to 4-ha watershed for each of the major management systems[a]

	1979			*1980*			*1981*		
Treatments	*I*	*O*	*O:I*	*I*	*O*	*O:I*	*I*	*O*	*O:I*
Traditional farming	3.6	8.4	2.4	33.7	125.6	3.7	3.7	16.9	4.6
Manual clearing, no-till	7.8	25.9	3.3	36.9	235.5	6.4	11.3	52.7	4.5
Manual clearing, conventional tillage	7.8	26.4	3.4	36.4	188.4	5.2	12.6	56.2	4.5
Shear blade clearing, no-till	3.7	32.9	8.9	36.9	219.8	6.0	11.3	66.5	5.9
Tree pusher/root rake, no tillage	3.0	23.0	7.7	36.9	312.4	8.5	11.3	55.3	4.9
Tree pusher/root rake, conventional tillage	3.0	29.1	9.7	36.4	282.6	7.8	12.6	36.8	2.9

[a]Inputs (I) and outputs (O) are expressed in GJ ha^{-1}.

farm income and its size will also grow over time. Some of the inputs that the farmer cannot now afford will become affordable and economical. Some of the capital-intensive inputs have to await the right stage of development. Until then we must make effort to get the ball rolling. This is not an "operation shoestring" but a positive approach to get the subsistence farmers walking before they can run.

The sequence of steps necessary to adopt an ecological approach to high and sustained production are shown in Figure 29. The important

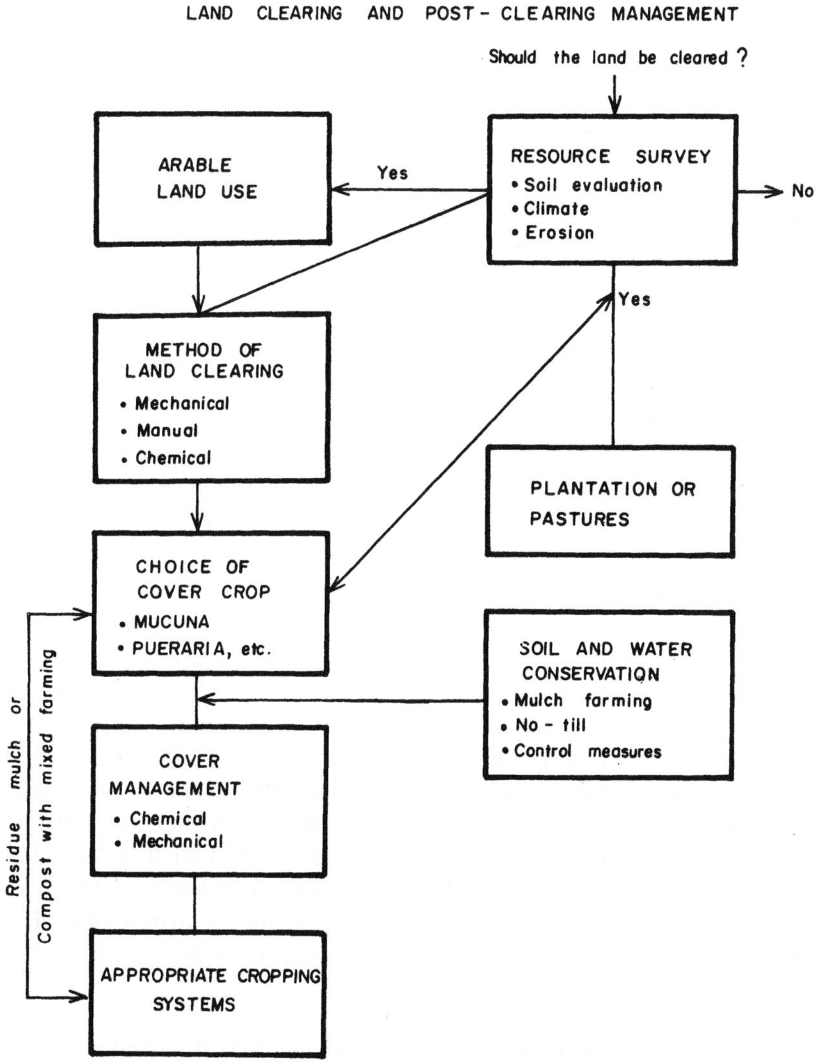

Figure 29. Sequence of steps needed for land clearing and postclearing soil management for high and sustained productivity.

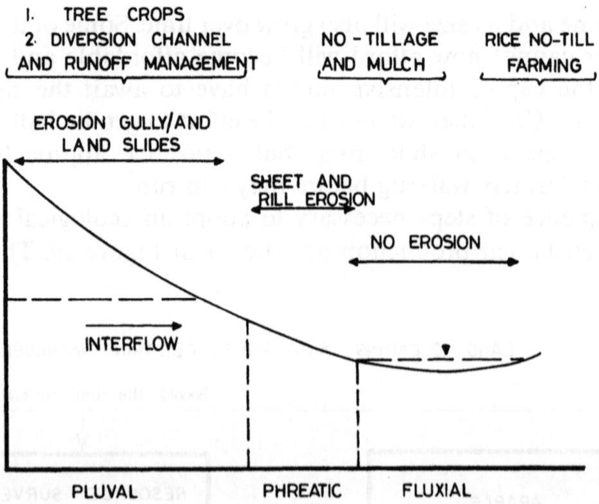

Figure 30. Adaptation of appropriate crops and soil management systems in relation to soil constraints (top), moisture regimen (bottom), and susceptibility of soil to erosion (center).

considerations are the choice of land-clearing methods and postclearing soil surface management systems that are ecologically compatible. Another important aspect is to adapt crop and farming systems according to the potential and restraints of the soils concerned (Figure 30). Within the management constraints of small landholders, soils can produce whatever they are inherently capable of doing.

IX. Research and Development Priorities

There is a need for developing a farming systems technology that enables people to live in comfort and in symbiosis with nature. There are still many misconceptions about the actual potential and constraints of tropical ecosystems. Although considerable progress has been made in the recent past toward replacing myths by facts, the reasons for the lack of sustained-yield farming systems in tropical ecosystems are not fully understood. The research and development objectives are to achieve high and sustainable yields but with low inputs and with less damage to soil and environments. The goal is to achieve optimum sustainable yields with modest inputs rather than the maximum yields based on high capital and energy inputs.

A. Improved Soil Management for Economically Viable Farming Systems

The first research priority should be to identify the reasons for the nonacceptability of the research information already available. Although the innovative concepts and subsystems discussed are technically sound, their economic evaluation and social acceptability need to be assessed for different socioeconomic environments. These concepts should be evaluated as integral parts of the overall system rather than as individual components of the improved technology. McCown (1984) rightly indicated that improved soil management is sustainable only within economically improved farming systems.

B. Integration of Components into Viable Farming Systems

The promising innovations and subsystems already developed to alleviate specific biophysical constraints related to soil and environments should be integrated into farming systems as specific case studies. Methodology involving linear programming and systems analysis should be standardized to facilitate establishment of blueprints of farming systems for local specific situations on the basis of subsystems and component technology already developed.

C. Resource Base

Baseline data are needed on soil, climate, water, and vegetation resources of the tropics, and their potentials and constraints. Resource data banks must be established for major agroecological zones, and their productive potential must be defined on the basis of conceptual models.

D. Soil Management

Rapid progress is expected within the next 10 to 15 years when the demands made on soil resources will be greater than ever before. The low-input systems now being recommended will be obsolete, and with an improved economic situation farmers will be eager to invest in agriculture as a commercial enterprise. When that happens, researchers should be ready to provide the components or subsystems of medium- to high-input technology. A specific research priority for soil resource management for the 21st century is: basic data from well-designed, long-term field experiments on soil management involving technology with different levels of inputs. The following research topics need to be considered for soils of the lowland tropics:

1. Land clearing and development in the forest region is still a major problem. With increasing labor constraints in the future, mechanized methods of land clearing will be increasingly relied upon. It is therefore important to develop systems of land clearing and management that will minimize the degradation effects of mechanized land clearing. Development of environmentally acceptable equip-

ment is essential since suitable land-clearing techniques will differ greatly among land types, soils, vegetation, and moisture regime.

2. Restoration of eroded and degraded lands deserves a high priority particularly to reduce the need to clear and develop new lands. Methods should be developed to restore and rehabilitate eroded and degraded lands. Land evaluation criteria should indicate the time when soil should be taken out of production and put under a restorative and ameliorative phase. Knowing the critical limits of soil properties for different levels of management is crucial in this endeavor.
3. The problem of soil compaction will become severe with intensive land use and increasing mechanization. There is a need to develop routine methods for characterizing soil compaction. There are not standardized procedures available to minimize soil compaction by motorized farm equipment, or to restore production on soils hitherto compacted. The problem should be addressed from various aspects, e.g., machinery, rotations, and farming systems. Traffic-induced soil compaction deserves an interdisciplinary research approach by soil and crop scientists working together with soil engineers and machinery designers.
4. More basic data are needed on soil erosion and its control and for predicting water runoff and soil loss under different land uses and in different farming systems. In addition to considering basic factors affecting erosivity and erodibility, there is a need to establish numerical limits of "soil loss tolerance." This information is important for long-range planning for different land uses. Also needed are the conceptual and empirical models relating crop yield and soil loss to different levels of management to assess the economic consequences of accelerated soil erosion. Unless the relationship between erosion and productivity is developed, it is difficult to plan development strategies and choose soil management methods.
5. Management of soil structure is still a mystery. Why does the structure of most tropical soils deteriorate so rapidly and how can this be prevented? Development of crust and surface seal is a major problem in soils containing predominantly low-activity clays, including those in the semiarid tropics. Techniques of soil surface management should address this problem.
6. The riddle of increasing agricultural output per unit input will remain a challenge for generations to come. The inputs may be natural resources or soil amendments. In the semiarid and arid tropics, increasing output per unit of water is the basic ingredient of a successful technology. For the humid regions with highly leached soils, limiting plant nutrients (e.g., P, N, lime) are the major consideration for developing improved farming systems. Also to be studied are soil–water–fertility interactions for different agroclimatic

considerations, including leaching patterns, and salt and water balance.

7. The development of appropriate tillage systems for different sized farms and for a wide range of soils, crops, and climatic environments is an important research consideration. Considering its merits in erosion control and water conservation, the potential of the no-till system should be exploited. The ecological limits for application of the no-till technique can be greatly extended by improving the package of agronomic practices required for its successful implementation. Tillage systems should be geared toward alleviating specific soil-related constraints to crop production, e.g., soil temperature, soil and water conservation, soil compaction, maintenance of soil structure, and soil organic matter content.
8. Study of soil dynamics or evolution of physical, chemical, and biological properties with time in some representative farming systems and varied land uses is needed to establish the cause–effect relationship between land use and soil properties. The agronomic output should always be assessed in terms of the impact of land use on soil properties.
9. The soils of humid regions have special toxicity constraints. Research is needed for crop adaptation to these nutrient constraints. In arid and semiarid ecologies, on the other hand, there is a problem of salt accumulation in the surface soil horizon. The basic studies of salt and water balance in different farming systems and in different ecologies should provide the basis of management systems to overcome these problems.
10. Integrating perennial crops and livestock with food crop production deserves to be given a top priority, especially in regions where rapid soil degradation is a severe problem with intensive arable land use.

References

ACIAR. 1984. Australian Centre for International Agricultural Research. Proc. Int. Workshop on Soils, Townsville, Old., Australia, Sept. 12–16, 1983.

Agboola, A. A. 1981. The effects of different soil tillage and management practices on the physical and chemical properties of soil and maize yield in a rain forest zone of Western Nigeria. *Agron. J.* 73:247–251.

Aina, P.O. 1979. Soil changes resulting form long term management practices in Western Nigeria. *Soil Sci. Soc. Am. J.* 43:173–177.

Aina, P.O., R. Lal, and G.S. Taylor. 1977. Soil and crop management in relation to soil erosion in the rain forest of Western Nigeria. In G. R. Foster (ed.) *Soil Erosion Prediction and Control.* Soil Conserv. Soc. Am. Spec. Publ. Ankeny, Iowa. 21; 75–82.

Ajunwon, S.O., B.A. Olunuga, and R. Lal. 1978. Applicability of no-tillage techniques in the forest and savannah regions of Western Nigeria. *Nigerian J. Sci.* 12:247–286.

Akobundu, I.O. 1980. Live mulch: a new approach to weed control and crop production in the tropics. C. Wheelers and J. Holroyd (compilers) *Proc. 1980 Brit. Crop Protection Conf.* 377–380.

Akobundu, I.O. 1982. Live mulch crop production in the tropics. *World Crops 34,* 125–126, 144–145.

Allan, W. 1965. *The African husbandman.* Oliver and Boyd, Edinburgh, 505 pp.

Amezquita, E.C. 1981. A study of water regime of a soil during approach to field capacity and wilting point. Ph. D. Disst., Univ. Reading, Reading, U.K., 244 pp.

Arledge, J.E. 1980. Soil conservation at work: Guatamala's small farmer project. *J. Soil Water Cons.* 35:179–189.

Ardnt, W., L.T. Phillips, and M.J.T. Norman. 1963. Comparative performance of crops on three soils of the Tipperary region. Tech. Paper 23, N.T. CSIRO, Aust. Div. Land Res. Reg. Serv. Catharine, Australia.

Babalola, O., and R. Lal. 1977. Subsoil gravel horizon and maize root growth I and II. *Plant Soil* 46:337–357.

Baconguis, S.R., and C.S. Rondilla. 1979. Inflitration studies on giant ipil-ipil (*Leucaena leucocephala*) plantation sites at the Buhisan watershed. *Sylvatrop* 4:23–29.

Baena, A.R.C. 1977. The effects of pasture on the chemical composition of the soil after cleanup and burning a typical tropicl highland rain forest.M.Sc Thesis, Iowa State University, Ames, Iowa, pp. 172.

Baffoe-Bonnie, E., and C. Quansah. 1975. The effect of tillage on soil and water loss. *Ghana J. Agr. Sci.* 8:191–195.

Baldanzi, G. 1959. Efeitos da Queimada sobre A Fartilidade do sol. Pelotas, Rio Granda do sul, Instituto Agronomico du Sul. In: F. Morgan (ed.), *Developing the Amazon. Indiana Univ. Press, Bloomington, 292 pp.*

Barney, G.O. 1980. Global 2000 report to the President of the USA. Pergamon Press, New York, 360 pp.

Baylis-Smith, T.P. 1982. *The ecology of agricultural systems.* Cambridge Univ. Press, Cambridge, pp. 112.

Benoit, J. 1977. The start of the growing season in Northern Nigeria. *Agr. Meteoral.* 18:91–99.

Bois, J.F., and Ph. Couchat. 1983. Comparison of the effect of water stress on the root systems of two cultivars of upland rice. *Ann. Bot.* 52:479–487.

Bonsu, M., and R. Lal. 1983. Hydrological properties of soils in Western Nigeria: a comparison of field and laboratory methods. *Nigerian J. Soil Sci.* 3:101–117.

Bonsu, M., and H.B. Obeng. 1979. Effects of cultural practices on soil erosion and maize production in the semi-deciduous rain forest and forest-savannah transition zones of Ghana. In: R. Lal and D.J. Greenland (eds.), *Soil physical properties and crop production in the tropics.* J. Wiley and Sons, New York.

Bornemisza, M.R.E. 1968. Algunas propiedades de suelos de la region Amazonica del Estado de Maranhao, Brasil. *Turrialba* 13:34–44.

Brown, L.R., and E.C. Wolf. 1984. Soil erosion: quiet crisis in the World economy. *World Watch Paper 60, World Watch Institute, Washington, D.C., 49 pp.*

Burgos-Leon, W., G. Ganry, R. Nicou, J. L. Chopart, and Y. Dommergues. 1980. Un ces de fatigue des sols induite par la culture du sorgho. Agron. Trop. 35;319–334.
Buringh, P. 1979. *Introduction to the study of soils in tropical and subtropical regions.* PUDOC, Wageningen, The Netherlands. pp. 124.
Cadima, A.Z., and P. de T. Alvim. 1973. Algunos factores del suelo asociados con la productividad del cacao tero en Bahia, Brazil. *Rev. Theobroma* (Brasil) 3:13–26.
Cadima, A.Z., and Z.Z. Marcos. 1981. Caracteristicas fisico-hidricas do solo relacionadas com a absorcao de agua pelo cacaneiro. *Rev. Theobroma* 11:133–143.
Chakela, W.K. 1981. *Soil erosion and reservoir sedimentation in Lesotho.* Scandinavian Istitute of African Studies, Uppsala, Sweden, ISBN 91-7106-186-X, 150 pp.
Chang. J. 1968. The agricultural potential of the humid tropics. *Geogi. Rev.* 53-(July): 333–361.
Charreau, C. 1972. Problemes poses par l' utilisation agricole des sols tropicaux par des cultures ammelles. *Agron. Trop.* 27:905–929.
Charreau, C. 1977. Controversial points in dryland farming practices in semi-arid West Africa. G.H. Cannell (ed.) *Symp. Rainfed Agric. Semi-arid Regions, Univ. Of California, Riverside. 313–360.*
Charreau, C., and R. Nicou. 1971. L'amelioration du profil cultural dans les sols sableux et sablo argileux de lazone tropicale seche ouest Africaine et ses incidences agronomiques. Agron. Trop. 26:209–255, 565–631, 903–978, 1184–1247.
Chopart, J.L. 1981. Le travail du sol au Senegal. Institut Senegalais De Recherches Agricoles (ISRA), Bambey, Senegal, 36 pp.
Chopart, J.L. and R. Nicou. 1976. Influence du labor sur les developpement radiculaire de defferentes plantes cultivees au Senegal—consequences sur leur alimentation hydrigue. *Agron. Trop.* 31:7–28.
Chopart, J.L., J.M. Kalms, J. Marquette, and R. Nicou. 1981. Comparison de differentes techniques de travail du sol en trois ecologies de l' Afrique de l' Quest. Institut De Recherches Agronomique Tropicales et Des Cultures Vivriers, Montpollier, France, 65 pp.
Cocheme, J., and G. Zazzara. 1969. A contribution to the forecasting of the early rains at Nairobi. *E. African Agr. Forest J.* 34;312–314.
Colmet-Daage, F., and F. Cucalon. 1967. Characteristiques de quelques sols d' Equateur derives de cendres volcaniques, Ier Partie: Essai de caracterisation. *Cah. ORSTOM Ser. Pedol.* 5:3–38.
Couper, D. C., R. Lal, and S. Claassen. 1981. Land clearing and development for agricultural purposes in Western Nigeria. In: R. Lal and Z.W. Russell (eds.), *Tropical Agricultural Hydrology.* J. Wiley and Sons, Chichester, U.K., pp. 119–130.
Couper, D. C., R. Lal, and S. L. Claassen. 1982. Agronomic problems with large scale mechanized farming. In: *Proc. R. Lal, P. A. Sanchez, and R. W. Cummings (eds.), Int. Conf. land Clearing Dev.,* A. A. Balkeme, Rotterdam (in press).
Critchley, B.R., A.G. Cook, U. Critchley T.J. Perfect, A. Russell-Smith, and R. Yeadon. 1979. Effects of bush clearing and soil cultivation on the invertebrate fauna of a forest soil in the humid tropics. *Pedobiologia* 19:425–438.
Dabin, B., and R. Maignien. 1979. Les principaux sols d' Afrique de l' Quest et leurs potentialites agricoles. *Cah. ORSTOM Ser. Pedol. XVII: 235–257.*

Dagg, M., and J.C. Macartney. 1968. The agromonic efficiency of N.I.A.E. mechanized tie ridge system of cultivation. Exp. Agr. 4:279–294.

Daniel, J.G., and A Kilasingam. 1975. Problems arising from large scale forest clearing for agricultural use: The Malaysian experience. *Planter (Malaysia)* 51:250–257.

Davies, H.R.J. 1973. *Tropical Africa, An Atlas for Rural Development.* Univ. Of Wales Press, Cardiff. *XIV* 81 pp.

De Martonne, E. 1926. *Geographie Phsique, III, Biogeographie.* 3rd ed. A. Colin, Paris.

Dhruva Narayana, V.V., and G. Sastry. 1985. Soil Conservation in India In: S.A.El-Sovaify ed-al (ed.), Soil Erosion and conservation. SCSA, Ankeny, Iowa; pp. 3–9.

Dowker, D.M. 1971. A note on the reduction in yield of Taboran maize by late planting. *E. Afr. Agr. Forest J.* 3:33–34.

Dregne, H.E. 1983. Desertification: Man's abuse of the land. *J. Soil Water Cons.* 33(1):11–14.

Dregne, H.E. 1983. Soils of semi-arid regions. In: E. Campos-lopez and R.J. Anderson (eds.), *Natural Resources and Development in Arid Regions.* Westview Press, Boulder, Colorado pp. 53–62.

Driven, J.G.P. 1977. Tropical grassland management based on ecological principles. *De Surinaamse Landbow* 25:1–8.

Dunham, R.J. 1982a. No-tillage crop production research at Samauru. Inst. for Agri. Research. Samaru, Zaria, Nigeria, 23 pp.

Dunham, R.J. 1982b. *Soil management research in the Nigerian savannah.* Institute of Agr. Research Samaru, Zaria, Nigeria, 33 pp.

Dunham, R.J., and A.J. Aremu. 1979. Soil conditions under conventional and zero tillage at Samaru. Soil Sci. Soc. Nigeria Conference, 1-6 March, Nigeria.

Duque, F.F., J.C. de Melo, R.L.P. de Sousa, and R.L. Gomide. 1980. In: D. Marchetti, and A.D. machado (eds.) *Cerrado, Uno e Manejo.* Editerra, Bresilia, Brazil, pp. 501–519.

El-Swaify, S.A., T.S. Walker, and S.M. Virmani. 1984. Dryland Management Alternatives and Research Needs for Alfisols in the Semi-Arid Tropics. *Consultants' Workshop,* Dec. 1–3, 1983, ICRISAT, Hyderabad, India, 34 pp.

FAO. 1981. *Agriculture: Towards 2000.* FAO, Rome; pp. 134.

FAO. 1984. *Land, Food and People.* FAO, Rome; 96 pp.

Felipe-Morales, C., C. Alegre, R. Meyer, and D. Berrios. 1978. Perdidas de agua, suelo y nutrientes bajo diversos sistemas de cultivo en la localidad de sau Ramon-chauchamayo, Duraute la compana Agricula 1976–1977. W.A. Wunche ex al (ed.) ll Encontro Nacional de Pasauisa sobre conservacao Du solo, Embrapa, Passo Fundo, April 22–28, 1978, pp. 311–321.

Fitzpatrick, E.A. 1965. Climate of the Tipperary Area. *General report on lands of the Tipperary area, 1961.* CSIRO, Melbourne, Australia, 112 pp.

Fitzpatrick, E.A., R.O. Slatyer, and A.I. Krishnan. 1967. Incidence and duration of periods of plant growth in central Australia as estimated from climatic data. *Agric. Meteorol.* 4:389–404.

Fouriner, F. 1963. The soils of Africa. In: *A review of the natural resources of the African continent.* UNESCO, Paris, France, pp. 221–248.

Freeman, P. 1980. State of the environment and natural resources. USAID reports, JRB Associates, Virginia, USA.

Ghildyal, B.P. 1978. Effects of compaction and puddling on soil physical properties and rice growth. In: *Rice and Soils,* IRRI, Manila, Phillipines, pp. 317–336.

Ghildyal, B.P., and T. Satyanarayana. 1969. Influence of soil compaction on shoot and root growth of rice (*Oryza sativa). Indian J. Agron.* 14(2): 187–192.

Ghuman, B.S., and R. Lal. 1983. Effect of crop cover on temperature regime of an Alfisol in the tropics. *Agron. J.* 75:931–936.

Golley, F.B. 1983. Ecodevelopment. IU: F.B. Golley (ed.), Tropical Rainforest Ecosystmes. Elsevier Scientific Publishing Co., Amsterdam, Netherland, pp. 335–344.

Greenland, D.J. 1975. Bringing the green revolution to the shifting cultivators. *Science* 190:841–844.

Hall, A.E., and C. Dancette. 1978. Analysis fallow-farming systems in semi-arid Africa using a model to simulate the hydrologic budget. *Agron J.* 70:816–823.

Harrison, P.H. 1984. Population, climate and future food supply. *Ambio* 13:161–167.

Honisch, O. 1974. Water conservation in 3 grain crops in the Zambezi valley. *Exp. Agr.* 10:1–8.

Huat, T.E. 1974. Effects of simulated erosion on performance of maize grown on serdang colluvium. Soil Cons. Rec. Rep. No. 1; Kuala Lumpur, Ministry of Agriculture and Fisheries. Mimeo.

Hudson, N.W. 1976. *Soil Erosion.* B.T. Batsford Ltd., Scarborough, Yorkshire, U.K. pp. 320.

Hulugalle, N.R., R. Lal, and C.H.H. ter Kuile. 1984. Soil physcial changes and crop root growth following different methods of land clearing in Western Nigeria. *Soil Sci.* 138:172–179.

Huxley, P.A. 1979. Zero-tillage at Morogoro, Tanzania. In: R. Lal (ed.), *Soil tillage and Crop Production. IITA Proc. Series 2, IITA, Ibadan, Nigeria, pp. 259–270.*

ICRISAT. 1978–79. Annual Report. ICRISAT, 1979, Hyderabad, India, 172–179.

IITA. 1981. Role of tied ridges in maize production in Upper Volta, *Research Highlights 1981,* International Inst. Trop. Agnc. Ibadan, Nigeria, pp. 7–10.

IITA. 1983. Effect of in-situ mulch on soil temperature and maize production. Research Highlights 1983. Int. Inst. Trop. Agnc., Ibadan, Nigeria, pp. 73–76.

IITA. 1984. Evaluating conservation-effective cropping systems for an African Savanna soil. Research Highlights 1984., Int. Inst. Trop. Agnc. Ibadan, Nigeria, pp. 103–107.

Isbell, R.F. 1978. Soils of the tropics and sub-tropics. In: C.S. Andrews and E.J. Kamprath (eds.), *Mineral Nutrition of legumes in Tropical and Sub-Tropical Soils.* CSIRO, Melbourne, Australia; pp. 1–19.

Islam, M.A. 1983. Soil erosion and conservation in the hilly areas of Bangla Desh. *Symp. Pros. Malama Aina Conference,* January 16–22, 1983, Honolulu, Hawaii.

Jenkins, D.S., and A. Ayanaba. 1977. Decomposition of carbon-14 labelled plant material under tropical conditions. *Soil Sci. Soc. Am. J.* 41:912–916.

Jones, M.J., and J. Sinclair. 1986. The effects of bare fallowing, previous crop and time of ploughing on soil moisture conservation in Botswana. *Field Crops Res.* (In Press).

Juo, A.S.R., and R. Lal. 1978. Nutrient profile in a tropical Alfisol under conventional and no-till systems. *Soil Sci.* 127:168–173.

Jurion, F., and J. Henry. 1969. Can primitive farming be modernized? I.N.E.A.C. Hors Series, Yangam bi, Zaire pp. 457.

Kampen, J., J. Hari Krishna and P. Pathak. 1981. Rainy season cropping on deep

Vertisols in the semi-arid tropics: effects on hydrology and soil erosion. In: R. Lal and R.W. Russell (eds.), *Tropical Agricultural Hydrology*. J. Wiley and Sons, Chichester, pp. 257–272.

Kang, B.T., G.F. Wilson, and L. Sipkens. 1981. Alley cropping maize with Leucaena in Southern Nigeria. *Plant Soil* 63:165–179.

Kannegieter, A. 1969. The combination of a short term pueraria fallow, zero cultivation and fertilizer application: Its effects on a following maize crop. *Trop. Agr.* 125:1–18.

Kar, S., and S.B. Varade. 1972. Influence of mechanical impedance on rice seedling root growth. *Agron. J.* 64:80–81.

Kellman, M.C. 1969. Some environmental components of shifting cultivation in upland Mindanao. *J. Trop. Geogr.* 28:40–56.

Kenworthy, J.M., and J. Glover. 1958. The reliability of the main rains in Kenya. *E. Afr. Agr. J.* 23:267–272.

Kemper, B., and R. Derpsch. 1981. Results of studies made in 1978 and 1979 to control erosion by cover crops and no-tillage techniques in Parana, Brazil. *Soil Till. Res* 1:253–267.

Khatibu, A.I., R. Lal, and R.K. Jana. 1984. Effects of tillage methods and mulching on erosion and physical properties of a sandy clay loam in an equatorial warm humid region. *Field Crops Res.* 8:239–254.

Kiley-Worthingon, M. 1981. Ecological argiculture, What it is and how it works. *Agr. Environ.* 6:349–381.

King, K.F.S. 1979. Agroforestry and the utilization of fragile ecosystems. *Forest Ecol. Manag.* 2:161–168.

Knorr, D. (ed.) 1983. *Sustainable Food Systems.* Ellis Horwood Limited, Publishers, Chichester, U.K. 416 pp.

Kowal, J.M. 1970. Some physical properties of soils at Samaru, Zaria, Nigeria, II Water storage characteristics. Res. Bull. Samaru 118, 52 pp.

Kowal, J.M., and A.H. Kassam. 1978. *Agricultural Ecology of Savanna.* Clarendon Press, Oxford, 403 pp.

Lal, R. 1973. Effects of seedbed preparation and time of planting maize in Western Nigeria. *Exp. Agri.* 9:303–313.

Lal, R. 1975. Role of mulching techniques in tropical soil and water management. IITA Tech. Bull. 1, 38 pp.

Lal, R. 1976. *Soil erosion problems on an Alfisol in Western Nigeria and their control.* IITA Monograph 1, 208 pp.

Lal, R. 1978. Influence of within- and between-row mulching on soil temperature, soil moisture, root development and yield of maize in a tropical soil. *Field Crops Res.* 1:127–139.

Lal, R. 1979. Physical characteristics of soils of the tropics: determination and management. In: R. Lal and D.J. Greenland (eds.), *Soil Physical Properties and Crop Production in the Tropics.* J. Wiley and Sons, Chichester, pp. 7–46.

Lal., R. 1981a. Deforestation and hydrological problems. In: R. Lal and E. W. Russell (eds.), *Tropical Agricultural Hydrology.* J. Wiley and Sons, Chichester, pp. 131–140.

Lal, R. 1981c. Soil erosion problems on an Alfisol in Western Nigeria and their control VI. Effects of erosion on experimental plots. *Geoderma* 25:215–230.

Lal, R. 1981. Clearing a tropical forest II. Effects on crop performance. *Field Crops Res.* 4:345–354.

Lal, R. 1983a. Erosion caused productivity decline in soils of the humid tropics. *Soil Tax. News* 5:4–5, 18.

Lal, R. 1983b. *No-till farming.* IITA Monograph 2, Ibadan, Nigeria, 64 pp.

Lal, R. 1985. Soil erosion and its relation to productivity in tropical soils. In: S.A. El-Swaify et al (ed.), Soil Erosion and Conservation. SCSA, Ankeny, Iowa, pp. 237–247.

Lal, R. 1984. Mechanized tillage systems effects on soil erosion from an Alfisol in watersheds cropped to maize. *Soil Till. Res.* 4:349–360.

Lal, R. 1985. Mechanised tillage systems effects on physical properties of an Alfisol in watersheds cropped to maize. *Soil Till. Res.* 6:149–61.

Lal, R., and D.J. Cummings. 1979. Clearing a tropical forest I. Effects on soil and microclimate. *Field Crops Res.* 2:91–107.

Lal, R. and B.T. Kang. 1982. Management of organic matter in soils of the tropics and sub-tropics. *Proc. 12th Int. Soc. Soil Sci.,* New Delhi 3:152–178.

Lal, R., G.F. Wilson, and B.N. Okigbo. 1978. No-tillage farming after various grasses and leguminous cover crops in tropical Alfisols I. Crop performance. *Field Crops Res.* 1;71–84.

Lal, R. G.F. Wilson, and B.N. Okigbo. 1979. Changes in properties of an Alfisol produced by various crop covers. *Soil Sci.* 127:377–382.

Lal, R., D. De Vleeschauwer, and R.M. Nganje. 1980. Changes in properties of a newly cleared Alfisol as affected by mulching. *Soil Sci. Soc. J.* 44:827–833.

Le Houerou, H.N., and G.F. Popov. 1981. An ecoclimatic classification of intertropical Africa. FAO Plant Production and Protection Paper 31, FAO, Rome, 40 pp.

Macartney, J.C., P.J. Nothwood, M. Dagg, and R. Dawson. 1971. The effect of different cultivation techniques on soil moisture conservation and the establishment and yield of maize at Kongwa, Central Tanzania. *Trop. Agr.* 48;9–17.

Machado, J.A. 1976. Efeito dos sistemes de cultivco reduzido e convencional ne alteracao de algumas propriededed fisicas e quimicas do solo. *Tese de livre doc.* UFSMIR. Mimeo, Res. Report.

Maduakor, H.O., R. Lal, and O.A. Opara-Nadi. 1984. Effects of methods of seedbed preparation and mulching on the growth and yield of white yam (*Dioscorea rotundata*) on an Ultisol in South-East Nigeria. *Field Crops Res.* 9: 119–130.

Maeda, T., and B.P. Warkentin. 1975. Void changes in Allophane soils determining water retention and transmission. *Soil Sci. Soc. Am. Proc.* 39:398–403.

Maeda, T, H. Takenaka, and B.P. Warkentin. 1977. Physical properties of Allophane soils. *Adv. Agron.* 29:229–269.

Maurya, P.R., and R. Lal. 1980. Effects of no-tillage and plowing on roots of maize and leguminous crops. *Exp. Agr.* 16:185–193.

Maurya, P.R., and R. Lal. 1981. Effects of different mulch materials on soil properties and on root growth and yield of maize and cowpea. *Field Crops Res.* 4:33–45.

Mbagwu, J.A.C., R. Lal, and T.W. Scott. 1983. Physical properties of three soils in Southern Nigeria. *Soil Sci.* 136:48–55.

Mbagwu, J.S.C., R. Lal, and T.W. Scott. 1984. Effects of artificial desurfacing on Alfisols and Ultisols in Southern Nigeria. Parts I and II. *Soil Sci. Soc. Am. J.* 48:823–838.

McCown, R.L. 1984. An agro-ecological approach to management of Alfisols in the semi-arid tropics. Div. of Tropical Crops and Pastures, CSIRO, Aitkenvale, Qld, Australia, Mimeo, 30 pp.

McCown, R.L., G. Haaland, and C. de Hann. 1979. The interaction between cultivation and livestock production in semi-arid Africa. In: A. R. Hall, G.H. Cannell, and H.W. Lawton (eds.), *Ecological Studies 34, Agriculture in Semi-Arid Environments.* Springer-Verlag, Berlin, pp. 297–332.

McCown, R.L., R.K. Jones, and D.C.I. Peake. 1980a. Alley farming system for the semi-arid tropics. In: I.M. Wood (ed.), *Proc. Aust. Agron. Cont.* Qld. Agr. Coll., Lawes, pp. 188–189.

McCown, R.L., R.K. Jones, and D.C.I. Peake. 1980b. Short term benefits of zero-tillage in tropical grain production. In: I.M. Wood (ed.), *Proc. Aust. Agron. Conf.* Qld. Agr. Coll. Lawes, pp. 220–221.

McCown, R.L., R.K. Jones, and D.C.I. Peake. 1985. Evaluation of a no-till tropical legume ley farming strategy. In: R. C. Muchow (ed.), *Agro-research for Australias semi-arid tropics,* Univ. Qld., Australin, pp. 450–472.

Melville, I.R. 1978. Conservation tillage for N.T. A look at minimum tillage systems in North America and Queensland. *N.T. Rural News-Magazine* 3 (4):14–15.

Monteith, J.L. 1972. Solar radiation and productivity in tropical ecosystems. *J. App. Ecol.* 9:747–766.

Mott, J., B.J. Bridge, and W. Arudt. 1979. Soil seals in tropical tall grass pastures of Northern Australia. *Aust. J. Soil Res.* 30:483–494.

Moureaux, C. 1959. Fixation de gaz carbonique par le sol. *Mem. Inst. Sci. Madagascar* D IX:109–120.

Myers, N. 1980. Conversion of tropical moist forests. National Academy of Sciences, Washington, D.C., pp.205.

Ngatunga, E.L.N., R. Lal, and A.P. Uriyo. 1984. Effects of surface management on runoff and soil erosion from some plots at Mlingano, Tanzania. *Geoderma* 33:1–12.

Nicou, R. 1974a. Contribution on the study and improvement of the porosity of sand and sandy-clay soils in the dry tropical zone. Agricultural consequences. *Agron. Trop.* 29:110–127.

Nicou, R. 1974b. The problem of caking with drying out of sandy and sandy-clay soils in the arid tropical zone. *Agron. Trop.* 30:325–343.

Nicou, R. 1977. Le travail du sol dans les terres exondees du Senegal: Motivations, contraintes. ISRA/CNRA, Bambey, Senegal, mimeo. 50 pp.

Nicou, R. 1979. Tillage in the West African tropical zone. Proc. 8th ISTRO Conf. International Soil Till Research Organization. Hohenheim, Germany, pp. 429–434.

Nicou, R., and C. Charreau. 1980. Mechanical impedance related to land preparation as a constraint to food production in the tropics (with special reference to fine sandy soils in West Africa). In: *Soil related constraints to food production in the tropics.* IRRI, Manila, Phillipines, pp. 371–388.

Norse, D. 1979. Natural resources development strategies and the world food problem. In: M.R. Biswas and A.K. Biswas (eds.), *Food, climate and Man,* Wiley Interscience, New York, pp. 12–51.

Nwa, E.U. 1977. Variability and error in rainfall over a small tropical watershed. *J. Hydrol.* 34:161–170.

Nye, P.H. 1955. Some soil forming processes in the humid tropics. IV The action of soil fauna. *J. Soil Sci.* 6:73–83.

Obeng, H.B. 1978. Soil, water, management and mechanization. *Afr. J. Agr. Sci.* 5:71–83.

Obi, A.O. 1974. The wilting point and available moisture in tropical forest soils of Nigeria. *Exp. Agr.* 10:305–312.

Ogunremi, L.T. 1983. Rice response to tillage and water regimes. Ph.D. Thesis, Univ. of Ibadan, Nigeria.

O'Keefe, P., and L. Kristoferson. 1984. The uncertain energy path and third World development. *Ambio* 13:168–170.

Okigbo, B.N. 1965. Effects of mulching and frequency of weeding on the performance of yield of maize. *Nigerian Agr. J.* 2:7–9.

Okigbo, B.N. 1978. Cropping systems and related research in Africa. AAASA Occasional Pub. Ser. OT-1, Addis Ababa, Ethiopia, 81 pp.

Okigbo, B.N., and R. Lal. 1980. Residue Mulches, intercropping and agri-silviculture potential in tropical Africa. In: S. Hill (ed.), *Basic Techniques in Ecological Farming IFOAM Conference Montreal, Canada, pp. 54–69.*

Olaniyan, G.O. 1983. No-tillage production of maize, rice, and cowpea in Nigeria. In: I.O. Akobundu and A.E. Deutsch (eds.), IPPC Document 46-B-83, Oregon State Univ., Corvallis, pp. 127–131.

Olsen, S.R., R.S. Watanabe, F.E. Clark, and W.D. Kemper. 1964. Effect of hexadecanol on evaporation of water from soil. Soil Sci. 97:13–18.

Othieno, C.O. 1975. Surface runoff and soil erosion on fields of young tea. *Trop. Agr.* 52;299–308.

Othieno, C.O., and D.H. Laycock. 1977. Factors affecting soil erosion under tea field. *Trop. Agr.* 54:323–331.

Panabokke, C.R. 1958. A pedologic study of dry zone soils. *Trop. Agr.* 114:151–174.

Pereira, H.C., E.M. Chenery, and W.R. Mills. 1954. The transient effects of grasses on the structure of tropical soils. *Emp. J. Exp. Agr.* 22:148–160.

Pereira, H.C., R.A. Wood, H.W. Brzostowski, and P.A. Hosegood. 1958. Water conservation by fallowing in semi-arid tropical East Africa. *Emp. J. Exp. Agr.* 26:203–228.

Pereira, H.C., P.H. Hosegood and M. Dagg. 1967. Effects of tied ridges, terraces and grass leys on a lateritic soil in Kenya. *Exp. Agr.* 3:89–98.

Perfect, T.J., A.G. Cook, B.R. Critchley, and A. Russell-Smith. 1981. The effect of crop protection with DDT on the microarthroped population of a cultivated forest soil in the sub-humid tropics. *Pedobiologia* 21:7–18.

Pidgeon, J.D. 1972. The measurement and prediction of available water capacity of ferrallitic soils in Uganda. *J. Soil Sci.* 23:431–444.

Pimental, D., and M. Pimental. 1979. *Food, Energy and Society.* Edward Arnold, Publishers Ltd., London, U.K.

Pla Sentis, I. 1977. *Metodologia para la caracterizacion fisica con fines de diagnostico de problemas de manejo y conservacion de suelos en condiciones tropiceles.* Instituto de Edafologia, Facultad de Agronomia, UCV, Maracay, Venezuela, 112 pp.

Plucknett, D.L., and N.J.H. Smith. 1982. Agricultural Research and Third World Food Production. *Science* 217:215–220.

Popenoe, H. 1960. Effects of shifting cultivation on natural soil constituents in Central America. Ph.D. dissertation, Gainesville, IFAS. Univ. Florida.

Portes, T. de Aquino, and C.H. Aidar. 1983. Temperatures, potenciais Hindricos do solo e perfis de luz no sistemes solteiro e de substituicao de feijao-milho. EMBRAPA-CNPAF Pesquisa en Andament No. 43, p. 5

Postel, S. 1984. Protecting forests. In: Brown, L. et al (ed.), *State of the World, 1984, A World watch Institute Report On Progress Toward a sustainable society.* W.W. Norton Co., New York, pp. 74–94.

Prentice, A.N. 1946. Tie-ridging, with special reference to semi-arid areas. *E. Afr. Agr. J.* 12:101–108.

Primavesi, A. 1982. *O Manejo Ecologico Do solo.* Nobel, Brazil, 541 pp.

Ramos, A.D., and H.E. Marinho. 1980. Erodibility of a lithosol without vegetation cover and under two vegetation types in the natives scrub pasture of Northern Brazil. *Boletim de Pesquisa* 2p 16 pp.

Rehm, S. 1978. Land development in the humid tropics. Agar-mechanisierung der DLG, 18–22, Frankfurt.

Reichart, K., G. Ranzani, E. De Freitas, Jr., and P.L. Libardi. 1980. Hydrology of some Amazonian soils. The lower Rio Negro region. *Acta Amazonica* 10:43–36.

Riehl, H. 1979. *Tropical Meteorology.* McGraw Hill, New York, 392 pp.

Rivas, J. DE. 1983. Erosion research in the Amazon jungle of Columbia *Symp. Proc. Malama Aina Conference* 16–22 January, 1983, Honolulu, Hawaii.

Roose, E.J. 1970. The relative importance of erosion and oblique and vertical drainage in the contemporary pedogenesis of ferrallitic soil in the Mid-Ivory Coast. *Cah. Pedo ORSTOM* 8:469–482.

Roose, E.J. 1975. Natural mulch or chemical conditioner for reducing soil erosion in humid tropical areas. Spec. Publ. Soil Sci. Soc. Am. 7;131–138.

Roose, E.J. 1977. Application of the USLE of Wischmeier and Smith in West Africa. In: D.J. Greenland and R. Lal. (eds.), *Soil conservation and management in the humid tropics.* J. Wiley and Sons, Chichester, pp. 177–188.

Rutherford, H. 1984. From shifting to semi-permanent and permanent farming in the African Savannas. *Int. Workshop on Farming Systems Proceedings.* ICRISAT, Hyderabad, India, pp. 325–350.

Sabel-Khoschella, U., R. Lal, and U. Schwertmann. 1983. Runoff and soil erosion studies in the savannah. *IITA Annual Report 1983,* pp. 148–149.

Sabel-Khoschella, U., R. Lal, and U. Schwertmann. 1983. Runoff and soil erosion studies in the savannah. *IITA Research Highlights* IITA Ibadan Nigen pp. 103–107.

Sanchez, P.A., and S.W. Buol. 1975. Soils of the tropics and the world food crisis. *Science* 188:598–603.

Sanchez, P.A., and J.G. Salinas. 1981. Low-input technology for managing Oxisols and Ultisols in Tropical America. *Adv. Agron.*. 34:279–406.

Sanchez, P.A., D.E. Brandy, J.H. Villachica, and J.J. Nicholaides. 1982. Amazon Basin soils: Management for continuous crop production. *Science* 216:821–827.

San Jose, J.J., and E. Medina. 1975. Effect of fire on organic matter production in a tropical savanna. In F.B. Golley and E. Medina (eds.), *Tropical ecological systems.* Springer-Verlag, Berlin, pp. 251–264.

Schöningh, E., and J. Alkämper. 1984. Effects of different mulch materials on soil properties and yield of maize and cowpea in an Eastern Amazon Oxisol. Proc.

First Symposium on the Humid Tropics, EMBRAPA, November 12–17, 1984, Belem, Brazil.
Schroo, H. 1976. Soil productivity factors of the soils of the Zandery formation in Surinam. *De Surinaamse Land.* 24:68–84.
Seguy, L., J. Kluthcouski, J.G. da Silva, F.N. Blumonscheim, and F.M. D-Acqua. 1984. Tecnicas de preparo do solo. EMBRAPA-CNPAF Circular Tecnica 17, 26 pp.
Seubert, C.E., P.A. Sanchez, and C. Valverde. 1977. Effects of land clearing methods on soil properties of an Ultisol and crop performance in the Amazon jungle of Peru. *Trop Agri.* (Trin.) 54;307–321.
Siband, P. 1972. Soil evolution under traditional cultivation in Haute-Casamance (Senegal): Main results. Agron. Trop. 27:574–591.
Sidiras, N., J.C. Henklain, and R.R. Derpsch. 1982. Comparison of three different tillage systems with respect to aggregate stability, the soil and water conservation and the yields of soybean and wheat on an Oxisol. Proc. 9th ISTRO Conf. ISTRO Osijek, Yogoslavia, pp. 537–544.
Silva, F.T.D., and M. Lourence. 1977. The importance of the utilization of bulldozers in the development of Cerrado soils in Brazil. *Agri. Sci. Bull.* 28:165–177.
Silva, L.F. da, and C.P. Pereira. 1979. Influencia dos horizontes do solo no solo no desenvolvimento de plantulas de cacau e na penetracao des suas saizes. *Rev. Theobrome* (Brasil) 7:13–18.
Silva, L.F. da. 1978. *Influencia do manejo de um ecossistema nes propriedades edaficas dos Oxisols de Tabuleiro.* Centro de Pesquisas do Cacau, CEPLAC, Itabuna, Bahia, Brazil.
Silva, L.F. da, and C.P. Perevia. 1979. Influencia dos horizontes do solo no desenvolvimento de plantules de cacau. *Rev. Theobroma* (Brasil) 9:85–92.
Silva, L.F. da. 1981. Edaphic changes in "tabuleiro" soils (Haploethox) as affected by clearing, burning, and management systems. Revista Theobromall pp. 5–19.
Sioli, H. 1973. Recent human activities in the Brazilian Amazon Region and their ecological effects. In: B. Meggers *et al.* (eds.), *Tropical forest ecosystems in Africa and South America. Smithsonian Institution, Washington, D.C. pp. 321–324.*
Soane, B.D. 1975. Studies on some soil physical properties in relation to cultivation and traffic. In: Soil physical conditions and crop production. Tech. Bull 29, Ministry of Agr., Fishries and Food, London, pp. 249–260.
Spratt, E.D., and S.L. Chowdhury. 1978. Improving cropping systems for rain-fed agriculture in India. *Field Crops Res.* 2:103–126.
Sreenivasan, P.S. 1971. A study of the district-wise distribution of rainfall in Bihar. *J. Hydrol.* 13:41–53.
Stibbe, E., and D. Ariel. 1970. No-tillage as compared to tillage practices in drylands farming of a semi-arid climate. *Neth. J. Agr. Sci.* 18:293–307.
Stone, L.F., and P.M. de Silva. 1978. Conductividade hidraulica de un latossolo Vermelho-Amarelo. *Pesq. Agropec. Bras. Brasilia* 4:63–71.
Stone, L.F., A.B. Dos Santos, and S. Steinmetz. 1980. Influencia de praticas culturais na capacidade de Retencao De Agua do Solo e no rendimento do Arrozde-Sequerio. *Pesq. Agropec. Bras. Brasilia* 15:63–68.
Suwardjo, S. Sukmaua, and N. Sofiyah. 1975. Beberapa data dan masalah

percobaan konservasi tanah nutuk pencegahan erosi. Departmen Pertanian Directorat Jendral Petaina, No. 11/1975, Bogor, Indonesia, 15 pp.

Swaminathan, M.S. 1982. Biotechnology research and third world agriculture. *Science 218:967–972.*

Takahashi, T., K. Nagahori, C. Mongkolasawat, and M. Cosiriku. 1983. Runoff and soil loss in shifting cultivation. In: K. Kyuma and C. Pairintra (eds.), Shifting cultivation. Ministry of Science and Technology, Bangkok, Thailand. pp. 84–109.

Tawonmas, D., M. Panichakul, S. Ratanarat, and W. Masangsul. 1984. Problem in Thailand. In: *Ecology and Management of Problem Soils in Asia.* FFTC Book Series No. 27, Taipei, Taiwan; pp. 50–57.

Tran-Vinh-An and H. Nguba. 1971. Contribution a l' etude de l' ean utile de quelques sols du Zaire. *Afri. Soils* 16:91–103.

Trewartha, G.T., and L.H. Horn. 1980. *An introduction to climate.* McGraw-Hill, New York, *XII*, 416 p.

UNESCO. 1978. *Tropical forest ecosystem: A state of knowledge report.* UNESCO, Paris.

Van der Graaff, R.H.M. 1965. In: *Soils of the Tipperary area.* Land Res. Ser. 13, CSIRO, Melbourne, Australia. pp. 68–80.

Van der Weert, R. 1974. The influence of mechanical forest clearing on soil conditions and resulting effects on root growth. *Trop. Agr.* 51:325–331.

Vergara, N.T. 1982. New directions in agro-forestry: The potential of tropical legume trees. E.P.J., East-West Centre, Honolulu, Hawaii, 36 pp.

Vine, P.N., R. Lal, and D. Payne. 1981. The influence of sands and gravels on root growth of maize seedlings. *Soil Sci.* 131:124–129.

Virgo, K.J., and R.N. Munro. 1978. Soil and erosion features of the Central Plateau region of Tigrai, Ethiopia. *Geoderma* 20:131–157.

Virmani, S.M. 1975. *The agricultural climate of the Hyderabad region in relation to crop planning.* ICRISAT, Hyderabad, India, mimeo.

Wall, J.R.D. 1981. A management plan for the Acellmate River Catchment, El Salvado: Soil conservation, river stabilization and water pollution control. *O.D.A. Land resource study* 30, 159 pp.

Walter, H. 1979. *Vegetation of the earth and ecological systems of the geobiosphere.* Springer-Verlag, New York, 274 pp.

Wang, S.T. 1984. Management of problem soils in Taiwan, ROC. In: *Ecology and Management of Problem Soils in Asia.* FFTC Book Series No. 27, Taipei, Taiwan; pp. 74–88.

Warnbecke, A. Van. 1978. Properties and Potentials of Soils in the Amazon Basin. *Interciencia* 3:233–242.

Whiteman, P.T.S. 1975. Moisture conservation by fallowing in Botswana. *Exp. Agr.* 11:305–314.

Wijewardene, R. 1981. Conservation farming. IITA/sri Lanka Program, Colombo, Sri Lanka, 19 pp.

Wilkinson, G.E. 1975. Effect of grass fallow rotations of Northern Nigeria. *Trop. Agr.* 52;97–103.

Wilkinson, G.F., and P.O. Aina. 1976. Infiltration of water into 2 Nigerian soils under secondary forests and subsequent arable cropping. *Geoderma* 15:51–59.

Willcocks, T.J. 1981. Tillage of clod-forming Sandy loam soils in the semi-arid climate of Botswana. *Soil and Till Res.* 1 (4):323–350.

Willey, R.W. 1979. Inter-cropping—its importance and research needs. *Field Crop Abs.* 32:2–10.

Wilson, G.F., R. Lal, and B.N. Okigbo. 1982. Effects of cover crops on soil structure and on yield of subsequent arable crops grown under strip tillage on an eroded Alfisol. *Soil Till. Res.* 2:233–250.

Wolf, J.M., and M. Drosdoff. 1976. Soil water studies in Oxisols and Ultisols of Puerto Rico II. Moisture retention and availability. *J. Agr. Univ. Puerto Rico.* 60:386–394.

Effect of Surface Residues on Soil Water Storage[1]

Darryl E. Smika and Paul W. Unger*

I. Introduction

The amount and availability of soil water are major factors influencing crop production. These factors influence crop and variety selection, planting time and method, fertilization, and weed control practices. Some soils may contain too much water and require drainage before crops can be successfully grown. For other soils, plant-available soil water may be limited and irrigation or water conservation practices may be needed for successful crop production. Management of crop residues to maintain them on the soil surface has generally increased water conservation. Three general factors affecting water conservation with surface residues are: (1) protecting the soil surface against raindrop impact energy to

[1]Contribution from USDA, Agricultural Research Service, Mountain Service, Mountain States Area and Southern Plains Area.

*Soil Scientists, Central Great Plains Research Station, Akron, CO 80720, and Conservation and Production Research Laboratory, Bushland, TX 79012.

Advances in Soil Science, Volume 5

maintain surface soil structure, thereby increasing infiltration and reducing runoff; (2) reducing evaporation; and (3) trapping and holding snow. The use of crop residues for water conservation has received more research attention in the arid and semiarid regions of the United States than in humid and subhumid regions. The arid and semiarid regions generally lie west of the 100th meridian, where annual precipitation is usually less than 500 mm, except for a narrow region of high rainfall immediately parallel to the west coast and in some intermountain valleys.

The value of stored soil water for crop production in arid and semiarid regions has long been recognized. Consequently, extensive research has been conducted since the early 1900s at numerous locations in arid and semiarid regions of the United States to develop practices for increasing the amount of water stored in soil from precipitation for subsequent use by crops. Most of the early research involved tillage methods, row spacings, fallowing, crop rotations, etc., and responses of various crops to these practices. The value of surface residues for conserving water was generally not considered until the late 1930s. Hallsted and Mathews (1936) stated that "trash and crop residues on the surface check runoff and allow more of the water to be absorbed by the soil," and some of the first studies involving residues were those by Duley and Kelly (1939), Duley and Russel (1939), and Russel (1939).

For this report, we review the early studies with surface residues. We then review the effects of surface residues on soil water storage as use of such residues has become practical in cropping systems, beginning with stubble mulch tillage in the 1940s, then progressing to the recent conservation tillage methods in which little or no tillage is performed. We also review the effects of surface residues on soil water evaporation and on trapping and holding snow.

II. Early Studies with Surface Residues

Duley and Kelly (1939) measured infiltration of simulated rainfall into soils ranging in surface texture from sandy loam to clay loam. The subsoils had textures ranging from sand to clay. The soils had either a bare cultivated surface or were covered with 5.6 metric tons per hectare of wheat straw (*Triticum aestivum* L.). For the six soils, total water infiltration averaged about five times greater, and final infiltration rate averaged about three times greater with the straw cover than with the bare surface. The differences in infiltration were attributed to the development of a very thin, dense surface layer on the bare soil. The straw prevented the formation of such a layer on the protected soil. They concluded that surface conditions had a greater effect on infiltration than soil type, slope, previous water content, and rainfall intensity. Although water contents were not reported, results of the study also suggested that more rainwater would be stored for subsequent use by plants in the straw-covered soil than in the bare soil.

The study by Duley and Russel (1939) involved straw (4.5 t ha^{-1}) on the soil surface, straw at the same rate disked or plowed into the soil, and several tillage treatments without straw. Effects of the treatments on soil water storage for the period from 23 April to 8 September 1938, are given in Table 1. Total rainfall for the period was 455 mm. The amount of water stored with the surface straw treatment was more than 2.5 times greater than that for the plowed treatment (no straw) and about two times greater than for the basin listing treatment. The latter treatment prevented runoff, yet less water was stored than with the straw treatment. This indicated that preventing runoff was not the total solution to conserving water in the U.S. Great Plains and that evaporation control afforded by surface residues was a major factor in soil water conservation. In several other experiments conducted by Duley and Russel (1939), soil water storage with straw on the surface was about 50 to 80 mm greater than that for a plowed treatment with straw.

Further evidence that reducing evaporation from the soil surface was a major factor in soil water conservation was provided by Russel (1939), who conducted field and soil column studies to determine the effect of surface residues on evaporation. Results of a field study are presented in Table 2. Water storage was greatest and evaporation was least where 17.9 t ha^{-1} of straw was on the surface. Evaporation loss was greatest with contour basin listing. These results indicated that reduced evaporation and favorable infiltration were important factors in improving soil water conservation.

Although the foregoing studies demonstrated the beneficial effects of

Table 1. Effect of straw and tillage treatments on soil water storage, 23 April to 9 September 1938, at Lincoln, Nebraska (adapted from Duley and Russel, 1939)

	Precipitation[a] stored		
Treatment	*mm*	*%*	*Depth of water penetration (m)*
Straw, 4.5 t ha^{-1}, on surface	247	54.3	1.8
Straw, 4.5 t ha^{-1}, disked in	176	38.7	1.5
Straw, 4.5 t ha^{-1}, plowed in	155	34.1	1.5
No straw, disked	89	19.6	1.2
No straw, plowed	94	20.7	1.2
Decayed straw, 2 t ha^{-1}, plowed in	79	17.4	1.2
Basin listed	126	27.7	1.5

[a]Precipitation totaled 455 mm.

Table 2. Water storage, runoff, and evaporation from field plots at Lincoln, Nebraska, 10 April to 27 September 1939

Treatment	*Storage (mm)*	*Runoff (mm)*	*Evaporation (mm)*	*Evaporative loss (%)*[a]
Straw, 2.2 t ha^{-1}, normal subtillage	30	26	265	83
Straw, 4.5 t ha^{-1}, normal subtillage	29	10	282	88
Straw, 4.5 t ha^{-1}, extra loose subtillage	54	5	262	82
Straw, 9.0 t ha^{-1}, normal subtillage	87	Trace	234	73
Straw, 17.9 t ha^{-1}, no tillage	139	0	182	57
Straw, 4.5 t ha^{-1}, disked in	27	28	266	83
No straw, disked	7	60	254	79
Contour basin listing	34	0	287	89

[a]Based on total precipitation, which was 321 mm for the period.

Adapted from Russel, 1939.

surface residues for conserving water, large-scale field operations with large amounts of crop residue on the surface were not practical at that time because suitable equipment for controlling weeds, maintaining crop residue on the soil surface, and planting crops under high-residue conditions had not yet been developed. However, those were the initial studies with stubble mulch tillage (subtillage), which subsequently was widely researched, especially in the U.S. Great Plains.

III. Stubble Mulch Tillage

Stubble mulch tillage was developed primarily for controlling wind erosion, but its value for reducing runoff and controlling water erosion was also soon apparent. The early research on stubble mulch farming summarized by McCalla and Army (1961), who concluded that stubble mulch farming was the most practical way to control wind erosion and that it reduced water erosion when adequate amounts of residue were present. However, additional soil conservation measures were needed to control water erosion effectively.

One factor contributing to the failure of stubble mulch tillage to control soil erosion by water is the relatively low amount of residue produced by rain-fed (dryland) crops in the Great Plains. Low residue production also contributed to the generally small increases in soil water storage with stubble mulch tillage as compared to clean tillage. Average soil water contents to a 1.5-m depth differed only 0.7% among stubble mulch, plow, and one-way disk treatments from 1941 to 1947 in Montana. The difference to a 1.8-m depth among similar treatments in Idaho was about 0.3%. In Kansas, available water contents to a 1.8-m depth were 135, 152, and 180 mm with one-way disk, plow, and stubble mulch tillage treatments, respectively. Subsequent wheat grain yields were 610, 450, and 930 kg ha^{-1}, respectively (McCalla and Army, 1961). At Bushland, Texas, soil water contents to a 1.8-m depth at wheat planting time were 12 and 27 mm greater with stubble mulch than with one-way disk tillage for continuous wheat and wheat–fallow systems, respectively (Figure 1). Respective grain yield increases with stubble mulch tillage were 110 and 130 kg ha^{-1} for the two systems (Johnson and Davis, 1972, Johnson *et al.*, 1974). For seven locations in the Central Great Plains, the average increase in soil water content with stubble mulch over clean tillage was about 25 mm (Smika, 1976a). Besides low residue production, other factors contributing to the low increases in water storage with stubble mulch tillage were the frequent tillage for timely weed control, which repeatedly exposed moist soil to the atmosphere and increased water loss to evaporation, and the failure of stubble mulch tillage to control all weeds effectively, which resulted in loss of some water via transpiration by weeds.

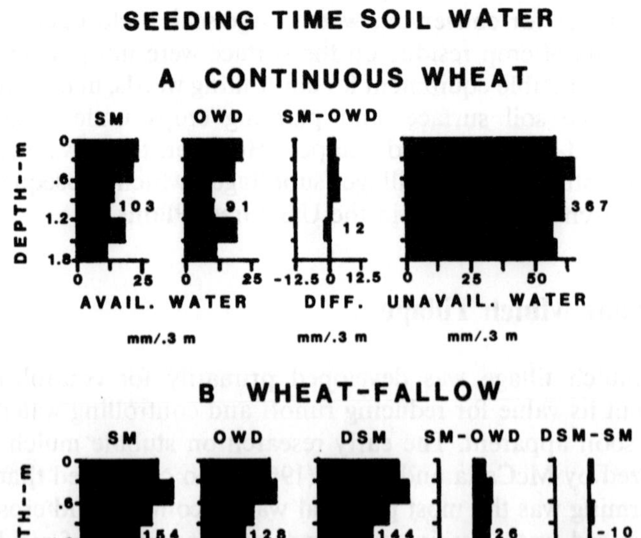

Figure 1. Average seeding-time soil water contents in continuous wheat (1942–1969) and wheat-fallow (1943–1969) plots at the USDA Conservation and Production Research Laboratory, Bushland, Texas. SM = stubble mulch, OWD = one-way disk, and DSM = delayed stubble mulch tillage (Johnson and Davis, 1972).

IV. Early Chemical Fallow (No Tillage)

When suitable chemicals (herbicides) became available, interest developed in substituting herbicides for tillage to control weeds (chemical fallow) and, thereby, to maintain more crop residues for a longer period of time on the soil surface to improve erosion control and water conservation. Although herbicides had previously been used on pastureland and in orchards (Wiese and Staniforth, 1973), the first experiments involving chemical fallow on cropland were initiated in Montana in 1948 (Baker *et al.*, 1956). Their study showed that where herbicides controlled the weeds, grain yields were comparable to those where conventional tillage was used for weed control. During the next 10 years, chemical fallow studies were initiated at several locations in the Great Plains. However, little progress was made with chemical fallow because the available herbicides did not kill all weeds during the fallow period (Wiese and Staniforth, 1973).

Another factor contributing to the lack of progress with chemical

Table 3. Straw mulch effects on soil water storage efficiency at Sidney, Montana; Akron, Colorado; and North Platte, Nebraska, from 1962 to 1965 (Greb *et al.*, 1967)

Mulch rate (t ha^{-1})	*Fallow period precipitation (mm)*	*Water storage efficiency (%)*
0	355	16
1.7	355–549	19–26
3.4	355–648	22–30
6.7	355–648	28–33
10.1	648	34

fallow was the fact that many of the studies involved dryland crops that produced relatively low amounts of residue. Although chemical fallow provided good protection against wind erosion, the limited amounts of residue were inadequate to improve water infiltration or suppress soil water evaporation. The limitation due to low amounts of residue was not recognized during the early studies but became apparent from studies by Greb *el al.* (1967) and Unger (1978), which showed that soil water storage during fallow progressively increased with increasing amounts of crop residues on the soil surface. Results of these studies are presented in Tables 3 and 4.

V. Chemical Fallow with Improved Herbicides

Although the early chemical fallow studies resulted in little progress, some progress occurred when atrazine [2-chloro-4-(ethylamino)-6-(iso-

Table 4. Straw mulch effects on soil water storage during fallow[a], water storage efficiency, and grain sorghum yield at Bushland, Texas, 1973–1976 (Unger, 1978)

Mulch rate (t ha^{-1})	*Water storage[b] (mm)*	*Storage efficiency[b] (%)*	*Grain yield (kg ha^{-1})*
1	72 c[c]	22.6 c[c]	1780 c[c]
1	99 b	31.1 b	2410 b
2	100 b	31.4 b	2600 b
4	116 b	36.5 b	2980 b
8	139 a	43.7 a	3680 a
12	147 a	46.2 a	3990 a

[a]Fallow duration of 10 to 11 months.

[b]Water storage determined to a 1.8-m depth. Precipitation averaged 318 mm.

[c]Column values followed by the same letter are not significantly different at the 5% level (Duncan's Multiple Range Test).

Table 5. Effect of fall weed-control treatments in new wheat stubble on soil water storage and grain yields in dryland rotations (Data from Greb, 1974, 1978)

Rotation and Fall weed control treatments	*Fall weed growth (kg ha^{-1})*	*Soil water storage at*		*Grain yield (kg ha^{-1})*
		Fall dormancy[a] *(mm)*	*Wheat or millet planting*[b] *(mm)*	
Wheat-fallow				
Akron, Colo. 1969–1972 (Greb, 1974)			Winter wheat	
Check, spring disk	1140	28	89	2420
Fall sweep, single	650	53	112	2690
Fall sweep, double	370	64	127	2940
Contact herbicide + atrazine + fall sweep	325	64	127	2940
Fallow-wheat-millet				
Akron, Colo. 1973–1977 (Greb, 1978)			Millet	
Check, spring disk	1175	1	114	1990
Fall sweep, double	505	19	133	2260
Contact herbicides + fall sweep	560	13	127	2210
Fall sweep + atrazine	390	17	135	2410

[a]From 25 October to 10 November.
[b]From 1 to 10 September for wheat and 1 to 10 June for millet.

propylamino)-*s*-triazine] and propazine [2-chloro-4,6-bis(isopropyl-amino)-*s*-triazine] herbicides became available for controlling weeds in cropping systems involving grain sorghum [*Sorghum bicolor* (L.) Moench]. A common cropping sequence in the Great Plains is winter wheat (10 months)-fallow (11 months)-sorghum (4 months)-fallow (11 months), whereby two crops are produced in 3 years.

Using the cropping sequence mentioned above, W.M. Phillips (1964) applied atrazine or propazine at 3.4 kg ha^{-1} to separate plots after wheat harvest, for weed control during the fallow period until grain sorghum was planted the next year. In some cases, 2,4-dichlorophenoxy acetic acid (2,4-D) at 1.1 kg ha^{-1} was also applied to control some existing weeds. The atrazine treatment gave excellent weed control. Propazine also gave good control, except for some grassy weeds that had emerged at the time of herbicide application. The cultivated plots required several tillage operations for weed control. The treatments did not differentially affect soil water storage, but sorghum on herbicide-treated plots yielded 4730 kg of grain per hectare and only 3040 kg ha^{-1} on cultivated plots.

After a few years, the weed population shifted from mainly broadleaf weeds that were susceptible to atrazine to grassy weeds [field sandbur, *Canchrus longispinus* (Hackel) Fern.] that were resistant, but grain sorghum yields were not significantly different from those with conventional tillage. When the atrazine-treated plots were tilled once at the time of atrazine application and again before planting or after sorghum emergence, no additional weed control measures were needed and the sorghum yielded 3720 kg of grain per hectare compared with 2290 kg ha^{-1} with conventional tillage (Phillips, 1969).

Although soil water contents were not reported for the above study, the higher grain sorghum yields with the atrazine–tillage (limited tillage) treatment were possibly attributed to higher soil water contents at planting and more efficient use of growing-season rainfall. Improved soil water storage and subsequently higher winter wheat and millet grain yields did result from various limited tillage treatments as illustrated by data (Table 5) from two studies in Colorado (Greb, 1974, 1978). The limited tillage treatments reduced weed growth, increased soil water storage, and increased grain yields compared with those of the check (spring disk) treatment. However, differences among the limited tillage treatments were generally small in most cases, indicating the importance of early (fall) weed control for conserving water and increasing crop yields.

Results from several other studies in the central Great Plains are presented in Tables 6, 7, and 8 (Smika and Wicks, 1968; Wicks and Smika, 1973). The results indicate that the gain in soil water was less for treatments involving plowing or disking, which incorporated surface residues, than for treatments involving stubble mulch or limited tillage or use of herbicides only. The greatest gain resulted from the herbicide

Table 6. Effect of tillage and herbicide treatments on soil water contents at the end of the fallow period[a] and on wheat and sorghum yields in a 3-year wheat–fallow–sorghum–fallow rotation (Smika and Wicks, 1968)

Treatment from wheat harvest to sorghum planting				*Grain yields*[d]	
Fall	*Spring*	*Treatment from sorghum harvest to wheat planting*[b]	*Soil water gain*[c,d] *(mm)*	*Wheat (kg ha^{-1})*	*Sorghum (kg ha^{-1})*
Subtillage	Disk	Subtillage (5)	186 b	3490 a	4080 b
Subtillage	Atrazine	Subtillage (4)	213 ab	3760 a	4200 b
Atrazine	Atrazine	Subtillage (4)	211 ab	3630 a	4580 ab
Atrazine	Atrazine	Contact herbicide (4–6)	223 a	3490 a	4890 a
Subtillage	Atrazine	Contact herbicide (4–6)	216 ab	3630 a	5020 a

[a]Fallow duration of about 11 months.

[b]Values in parentheses denote number of operations.

[c]Determined to a 3-m depth.

[d]Average values in a column followed by the same letter or letters are not significantly different at the 5% level (Duncan's Multiple Range Test).

Table 7. Effect of tillage and herbicide treatments on soil water contents at the end of the fallow period and on wheat yields in a 2-year (wheat–fallow) rotation (Smika and Wicks, 1968)

Operations during fallow[a]			
Initial operation following wheat harvest	*Subsequent operations*[b]	*Soil water gain*[c,d] *(mm)*	*Grain yield*[d] *(kg ha^{-1})*
Plow	Subtillage (5)[c]	186 c[d]	3090 b
Subtillage	Subtillage (5)	238 b	3360 ab
Atrazine followed by subtillage	Subtillage (5)	272 b	3290 ab
Atrazine	Subtillage (4)	275 b	3360 ab
Atrazine	Contact herbicides (4–6)	325 a	3560 a

[a]Fallow duration of about 14 months.

[b]Values in parentheses denote number of operations.

[c]Determined to a 3-m depth.

[d]Average values in a column followed by the same letter or letters are not significantly different at the 5% level (Duncan's Multiple Range Test).

Table 8. Effect of tillage and herbicide treatments on number of operations needed for weed control, residues maintained on the surface, soil water storage during fallow, and wheat yields in a 2-year (wheat–fallow) rotation

	Operations during fallow[a]				
Treatment	*Tillage*[b]	*Herbicide application*	*Residues maintained*[c] *(%)*	*Soil water gain*[d] *(mm)*	*Grain yield (kg ha*$^{-1}$*)*
Plow	8.5	0.0	0	146	2690
Stubble mulch	8.7	0.0	21	203	2880
Atrazine + stubble mulch	7.6	1.4	21	215	2910
Atrazine + contact herbicide + stubble mulch	5.1	2.8	25	237	3040
Atrazine + contact herbicide	0.0	6.0	46	274	3170

Reprinted with permission from Wicks and Smika, *J. Weed Sci. Soc. Am.*, Vol. 21, pp. 97–102. Copyright 1973 by the Weed Science Society of America.

[a]Fallow duration of about 14 months.

[b]The plow treatment included one moldboard plowing in the spring. Other tillage was with sweep implement.

[c]Average amount of residues at start of fallow was 6600 kg ha^{-1}.

[d]Determined to a 3-m depth.

treatments, which permitted a high percentage of residues to be maintained on the soil surface and no tillage to degrade surface soil structure to reduce water infiltration (Table 8). In general, grain yields increased with increases in stored soil water.

The foregoing examples have shown some major increases in water conservation and crop yields as a result of using various residue management practices compared with those obtained with clean tillage. The progressive improvement in water conservation and crop yields at Akron, Colorado, as influenced by changing management practices is illustrated in Table 9 (Greb, 1979). The approximate doubling of precipitation storage (soil water gain) and of grain yields is attributed to better weed control and to maintenance of residues on the surface, which enhances water infiltration and suppresses evaporation. However, the yield increases per unit increase in soil water content at planting are greater than those reported by Johnson (1964), which suggests that improved varieties, fertility, and weed control, as well as improved use of

Table 9. Progress in fallow systems with respect to water storage and wheat yields, Akron, Colorado (adapted from Greb, 1979)

		Fallow water storage		
Years	*Tillage during fallow*[a]	*mm*	*% of precip.*	*Wheat yield* ($kg\ ha^{-1}$)
1916–1930	Maximum tillage; plow, harrow (dust mulch)	102	19	1070
1931–1945	Conventional tillage; shallow disk, rod weeder	118	24	1160
1946–1960	Improved conventional tillage; begin stubble mulch in 1957	137	27	1730
1961–1975	Stubble mulch; begin minimum tillage with herbicides in 1969	157	33	2160
1976–1990	Projected estimate; minimum tillage; began no tillage in 1983	183	40	2690

[a]Based on 14-month's fallow, from mid-July to second mid-September.

growing season precipitation, probably contributed to the yield increases.

VI. Managing Irrigated Crop Residues

As previously stated, residue production by dryland crops may be too low in many cases to influence soil water storage markedly, even when most of the residues are maintained on the soil surface. This situation often occurs in the southern Great Plains. However, irrigated crops, such as winter wheat, produce relatively large amounts of residue. Consequently, limited- and no-tillage studies involving residues from irrigated winter wheat were initiated at Bushland, Texas in the late 1960s (Unger *et al.*, 1971). For the fallow period from wheat harvest in July 1968 until sorghum planting in May 1969, combinations of disk, sweep, and herbicide treatments were used to control weeds and volunteer wheat. The irrigated wheat produced 11 tons of residue per hectare. Water storage with treatments involving herbicides (atrazine and 2,4-D alone or after one sweep tillage) was about double the average water storage for all tillage-only treatments (disk, sweep, or disk plus sweep).

In 1970, winter wheat was planted after irrigated corn (*Zea mays* L.) without tillage or after rotovator tillage and on a conventionally tilled fallowed area, which resulted in areas of high, medium, and low residue levels, respectively (Unger and Parker, 1975). Within these treatments, one to five irrigations of the wheat resulted in different residue production levels. After wheat harvest, low-residue treatment plots were disked once to incorporate some residues with soil. All plots were treated with atrazine and 2,4-D to control weeds and volunteer wheat. Due to treatments imposed, surface residues at sorghum planting time (11 months after wheat harvest) ranged from about 0.2 to 6.7 t ha^{-1} and stored from 11 to 45% of the fallow period precipitation (467 mm) as soil water. The highest water storage, however, was not associated with the highest residue level, possibly because of either a higher initial soil water content or precipitation interception and absorption by residues. The dryland sorghum, planted without tillage, yielded from 2970 to 6010 kg of grain per hectare. The high yield was 2240 kg ha^{-1} greater than that of sorghum on an adjacent fallowed area where conventional production practices were used.

Unger and Wiese (1979) used no-, sweep-, and disk-tillage methods for managing wheat residues and controlling weeds during fallow in an irrigated winter wheat–fallow–dryland grain sorghum cropping system at Bushland, Texas. Soil water storage averaged 35.2, 22.7, and 15.2% of the precipitation; plant-available water to a 1.8-m depth at sorghum planting averaged 217, 170, and 152 mm; sorghum grain yields averaged 3140, 2500, and 1930 kg ha^{-1}; and water-use efficiencies for grain averaged 0.89,

0.77, and 0.66 kg m^{-3} for the treatments, respectively. All differences were statistically significant. In an irrigated winter wheat–dryland grain sorghum–dryland sunflower (*Helianthus annuus* L.) rotation study at Bushland (Unger, 1984), water storage during fallow after wheat and sorghum grain yields were highest with no tillage (Table 10). The tillage treatments applied during fallow after wheat had no residual effects on yields of sunflower and wheat that followed sorghum in the rotation.

In an irrigated winter wheat–fallow–irrigated grain sorghum study at Bushland, Texas, water storage during fallow averaged 35% of precipitation with no tillage and 21% with disk tillage (Musick *et al.*, 1977). Sorghum grain yields (1970–1971 average) were 5100 and 4080 kg ha^{-1} with no- and disk-tillage treatments, respectively, with 150 mm of irrigation during the growing season. With 300 mm of irrigation, the respective yields were 6460 and 5970 kg ha^{-1}. Because the plots were in level borders and no runoff occurred, differences in water storage during fallow resulted from lower evaporation with the no-tillage treatment. On graded-furrow plots, water storage during fallow averaged 47 and 28% of precipitation with no- and disk-tillage treatments, respectively. During the growing season, 169 mm of irrigation water infiltrated the soil on no-tillage plots, but only 89 mm on disk-tillage plots. The sorghum yielded 5420 and 4260 kg of grain per hectare with no- and disk-tillage treatments, respectively.

The above studies show that surface residues can have a major influence on infiltration of irrigation water as well as on storage of precipitation as soil water. The effects of surface residues on irrigation water infiltration, however, have been variable. For continuous sorghum, infiltration averaged 271 and 300 mm with conventional and no-tillage treatments, respectively. For continuous winter wheat, infiltration averaged 391, 353, and 356 mm with conventional-, limited-, and no-tillage treatments, respectively. Although residues from continuous wheat in the furrows retarded water flow and increased infiltration, loosening the surface soil by tillage also increased infiltration (Musick *et al.*, 1977). In another study with different tillage methods for continuous wheat production (Allen *et al.*, 1976), infiltration differences due to tillage were not significant for two crops, but in the third year, infiltration was higher with clean tillage because the soil was loosened by chiseling.

Sorghum residues in the furrows slowed irrigation water advance with no tillage as compared to that with clean tillage (Allen *et al.*, 1975). For four irrigations in 1971, infiltration was 323 mm with no tillage and 267 mm with clean tillage. A fifth irrigation totaling 82 mm was followed immediately by 63 mm of rainfall. Infiltration for the two events totaled 99 mm with no tillage and 87 mm with clean tillage. Sorghum grain yields, however, were similar for the two treatments, averaging 6180 and 6360 kg ha^{-1} with the no- and clean-tillage treatments, respectively. The no-tillage treatment resulted in an excessive number of volunteer

Table 10. Effect of tillage method on average soil water storage during fallow after irrigated winter wheat and on subsequent rain-fed grain sorghum yields at Bushland, Texas, 1978 to 1983 (Unger, 1984)

Tillage treatment	*Precipitation*		*Water storage*[b,c] *(%)*	*Grain yield*[c] *(kg ha^{-1})*
	Fallow[a] *(mm)*	*Growing season (mm)*		
Moldboard	316	301	29 b	2560 bc
Disk	316	301	34 ab	2370 cd
Rotary	316	301	27 b	2190 d
Sweep	316	301	36 ab	2770 b
No tillage	316	301	45 a	3340 a

[a]Fallow duration of 10 to 11 months.
[b]Based on fallow period precipitation stored as soil water.
[c]Column values followed by the same letter to letters are not significantly different at the 5% level based on Duncan's Multiple Range Test.

sorghum plants, which increased total dry matter yields but which adversely affected grain yields. In subsequent studies with continuous grain sorghum, Allen *et al.* (1980) have found that limited tillage effectively controls volunteer sorghum plants and that irrigation water infiltration and grain yields are higher with limited tillage than with clean tillage.

VII. Residue Effects—Subhumid and Humid Regions

Precipitation reliability is greater in subhumid and humid regions than in arid and semiarid regions. However, water deficiencies that adversely affect crop growth and yields do occur in subhumid and humid regions because of short-term droughts and low water-holding capacities of some soils (Griffith *et al.*, 1977; Reicosky *et al.*, 1977). Where these conditions prevail, additional water for crop use can be provided by increasing infiltration and reducing evaporation by using residue management practices. Subsurface tillage and chisel tillage used for stubble mulch system usually are less effective for increasing soil water storage in wetter regions than in drier regions because weeds often are not effectively controlled by these tillage methods in the wetter regions (McCalla and Army, 1961). Consequently, more intensive tillage methods have been widely used in more humid regions, such as the eastern Corn Belt (Griffith *et al.*, 1977). Intensive tillage, however, frequently leads to excessive soil erosion by water and practices to control this erosion have been evaluated at numerous locations.

To control erosion, runoff must be controlled by reducing the amount and the rate of runoff. Both factors are affected by surface residues because they enhance infiltration, which reduces runoff amount and retards flow across the surface, thus reducing runoff rate. Because infiltration is increased, more water is stored in the soil, provided the soil has capacity to store the additional water.

The amount of surface residues needed to control erosion has been widely researched. Mannering and Meyer (1963), for example, showed that straw mulch applications of 2.2, 4.5, and 9.0 t ha^{-1} to a silt loam on a 5% slope all maintained very high infiltration rates and resulted in essentially no erosion. Mulch rates of 0.6 and 1.1 t ha^{-1} resulted in 6.7 and 2.2 tons of soil loss per hectare, respectively, whereas bare soil lost 26.9 t ha^{-1}. The surface mulch increased the potential for recharging the soil profile with water by retarding runoff, thereby, increasing the opportunity for infiltration.

In recent years, much research has been conducted on conservation tillage systems in subhumid and humid regions. The residues maintained on the surface have resulted in higher soil water contents at many locations, as reported by Bennett (1977), Blevins *et al.* (1971), Jones *et al.*

(1969), Nelson *et al.* (1976) Mannering *et al.* (1975), Mannering and Meyer (1963), Van Doren *et al.* (1976), and others. The additional soil water resulting from maintenance of surface residues resulted in higher yields than with clean tillage, particularly when soil water became limited during the growing season.

VIII. Surface Residue Effects on Evaporation

In the foregoing sections, the increased soil water storage with surface residues resulted from increased infiltration and undoubtedly decreased evaporation. This was indicated by the early work of Duley and Russel (1939) and Russel (1939). The actual contribution of decreased evaporation on water conservation, however is difficult to determine in most field studies, but some studies have clearly shown that lower evaporation with surface residues is a major factor. Bond and Willis (1970) showed that small amounts of plant residue on the soil surface effectively decreased evaporation during the first stage of drying by increasing the resistance of water flow from the soil surface to the atmosphere. First, residues decrease surface temperatures, as shown in Table 11, which results in a decrease in vapor pressure of the soil water; second, the residue layer increases the thickness of the relatively nonturbulent air layer above the soil surface, which decreases vapor transport away from the soil. The latter was demonstrated by Smika (1983) under field conditions where the same amount of wheat residue on soil, but in different positions, resulted in different rates of water loss by evaporation (Figure 2). As the amount of standing straw increased, greater windspeed was needed for initial water loss and the rate of water loss also decreased with increasing amounts of

Table 11. Average daily soil surface temperature as affected by bare soil and wheat straw position during 5-week, August–September period, Akron, Colorado (Smika, 1983)

Straw position[a]	*Soil surface temperature*[b,c] *(°C)*
Bare soil	47.8 c
Flat straw	41.7 b
3/4 flat, 1/4 standing	39.6 b
1/2 flat, 1/2 standing	32.2 a

[a]All straw amounts were 4600 kg ha^{-1}.

[b]Average of measurements at 1000, 1200, and 1500 hr with a radiation thermometer.

[c]Values accompanied by different letters are significantly different at $P = 0.01$ (Duncan's Multiple Range Test).

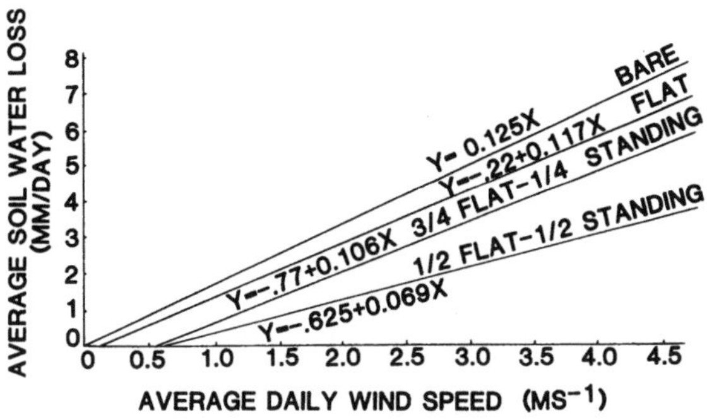

Figure 2. Effect of residue orientation and windspeed on soil water evaporation (Smika, 1983).

standing residue at a given windspeed. This clearly shows that reduced windspeed at the soil surface due to surface residues is a major factor in reducing evaporation, thereby, increasing soil water storage.

After the soil surface dries, water flow to the surface and porosity or air permeability of the surface soil become more important in the evaporation process. This was clearly demonstrated where conventional-, minimum-, and no-tillage treatments were compared for soil water loss during a 34-day period following 165 mm of rainfall and during which time no additional precipitation was received (Smika, 1976b). On the first day after the rain (Figure 3), there was very little difference in soil water content between treatments to a 15-cm depth. After 34 days (Figure 4), soil with the conventional-tillage treatment had dried to less than 0.1 cm cm^{-1} of water to a 12-cm depth and that of the minimum-tillage treatment had dried to that water content level to a 9-cm depth. Both depths are the same at which a blade tillage operation was performed 8 days before the rain. In contrast, soil with the no-tillage treatment dried to that water content level to only a 5-cm depth. For each treatment, some water loss occurred to soil depths greater than those indicated. At the time of this 34-day drying cycle, the conventional, minimum, and no-tillage treatments had 1200, 2200, and 2700 kg ha^{-1} of residue on the soil surface, respectively. The large amount of partly standing residue on the soil surface of the no-tillage treatment was sufficient to limit water loss to the 4-cm depth. However, where the residue amounts were less and the soil had been loosened by tillage, drying occurred to the depth of the tillage.

Decreased evaporation with surface residues is also a factor in greater water storage in more humid regions (Phillips, 1974). For a silt loam in Kentucky, evaporation from May through September was 2.4 times greater with conventional tillage than with no-tillage. This difference provided 18% more water for transpiration by no-tillage corn than for

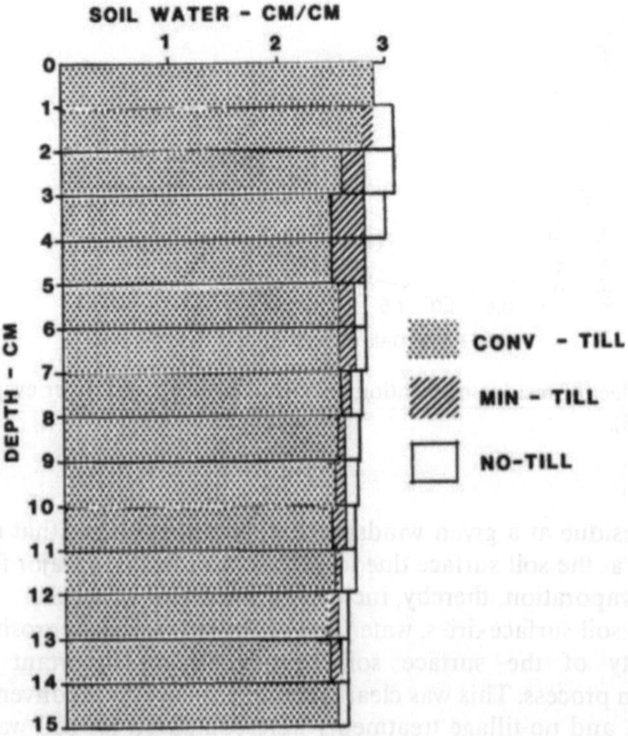

Figure 3. Soil water to a 15-cm depth on day 1 after 165 mm of rainfall as influenced by tillage treatment (Smika, 1976b).

conventional tillage corn (330 vs 280 mm of water during the growing season).

In contrast to field studies where evaporation measurements may be confounded by differences in infiltration, redistribution, and deep percolation due to differences in the amount of residues on the soil surface, the influence of surface residues on evaporation can be better investigated under controlled laboratory conditions. The studies by Bond and Willis (1969, 1970, 1971), which involved soil columns initially wetted to a given water content, showed that increasing amounts of surface residues decreased evaporation, especially during the first stage of evaporation. Because of lower evaporation during the initial stage, water from beneficial precipitation would move deeper into the soil where it was less subject to loss during the latter stages of evaporation. Although differences in evaporation rates between bare and residue-covered soil occur during the initial stage, evaporation rates for mulched soil may be greater during the later stages and, if continued long enough, evaporation from residue-covered soils may eventually approach, equal, or possibly

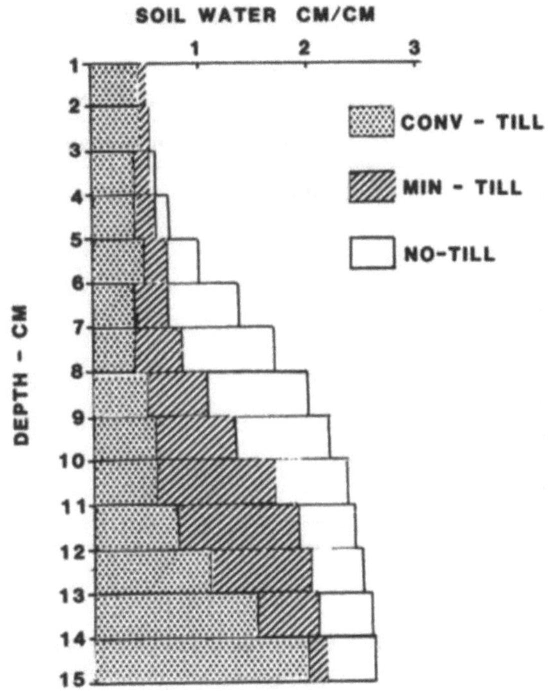

Figure 4. Soil water to a 15-cm depth 34 days after 165 mm of rainfall as influenced by tillage treatment (Smika, 1976b).

exceed that from bare soil. However, until that point is reached, the residue-covered soil will have more stored water than the bare soil that is potentially available for crop use.

Depth of water penetration was shown by Unger (1976) to be a major factor in reducing evaporation and in increasing soil water storage. When 15 mm of water was applied to a clay loam or a sandy loam at 14-day intervals, surface residue rates had little effect on the percentage of water stored. However, with 30-, 60-, or 120-mm water applications, the percentage of water stored in the soils progressively increased with increasing amounts of surface residues. For example, water stored in the clay loam with 30-, 60-, and 120-mm water additions was about 10 and 42%, 26 and 72%, and 48 and 84%, respectively, for the bare soil and 12 tons of residue per hectare treatments.

The foregoing examples in this section involved wheat residues applied at different rates on a surface area basis. Different types of residue have different specific gravities and, consequently, result in different percentages of surface coverage and residue thickness with a given amount of residue by weight. Unger and Parker (1976) used wheat, grain sorghum, and cotton (*Gossypium hirsutum*) (stalk) residues in an evaporation study

Table 12. Characteristics of residues used in an evaporation study (Unger and Parker, 1976)

	Residue type		
	Wheat	*Grain sorghum*	*Cotton*
Center of residues	Hollow	Pithy	Woody
Specific gravity	0.17	0.26	0.49
Thickness (cm) at:			
4 metric tons ha^{-1}	—	1.0	0.5
8 metric tons ha^{-1}	2.9	1.9	1.1
16 metric tons ha^{-1}	—	3.1	1.4
32 metric tons ha^{-1}	—	4.5	3.4
Surface coverage (%) at:			
4 metric tons ha^{-1}	—	66	8
8 metric tons ha^{-1}	100	90	37
16 metric tons ha^{-1}	—	98	80
32 metric tons ha^{-1}	—	100	99

in the laboratory. Characteristics of the residues with respect to specific gravity, thickness, and surface coverage at various rates are shown in Table 12. Effects of the residues at different rates on cumulative evaporation with time when the potential evaporation rate was 6.6 mm day^{-1} are shown in Figure 5. Because of the thickness and surface coverage differences, about 16 tons of sorghum and more than 32 tons of cotton residue per hectare were needed to obtain an evaporation

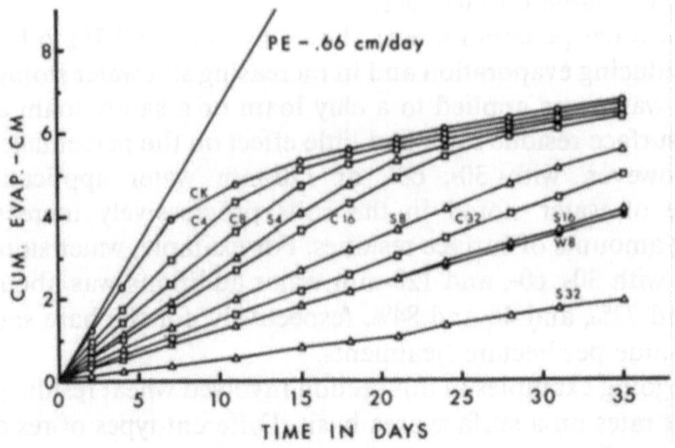

Figure 5. Effects of residue treatments on cumulative evaporation at 0.66 cm day^{-1} potential evaporation. CK = check, C = cotton, S = grain sorghum, W = wheat. Numbers after letters designate metric tons of applied residues per hectare. (Unger and Parker, 1976).

reduction comparable to that obtained with 8 t ha^{-1} of wheat residue. At potential evaporation rates of 9.2 and 12.9 mm day^{-1}, evaporation from the residue-covered soils was only slightly greater than when the potential rate was 6.6 mm day^{-1}. Other characteristics of residues that affect evaporation are rainfall interception; reflectivity, which affects the energy arriving at the soil surface; residue orientation (flat or matted vs standing), which affects thickness and porosity of the residue layer; uniformity of the surface residue layer; and aerodynamic roughness resulting from the residues (Van Doren and Allmaras, 1978).

IX. Surface Residue Effects on Water Conservation from Snow

The effectiveness of crop residues (mainly wheat) for increasing soil water storage during a fallow period increases from south to north. This increase is partially due to the effectiveness of standing residues for trapping and holding snow, providing the potential for increasing soil water storage when it melts.

The contribution of snow to water conservation is greater in the northern locations than in the southern locations because snow comprises a larger percentage of the annual precipitation at the northern locations. For example, water in snow accounts for about 15 to 20% of the average annual precipitation at locations from the central Great Plains in the United States to southern Canada (Saskatchewan) (De Jong and Steppuhn, 1983; Smika and Whitfield, 1966), whereas water in snow accounts for only about 5% of the annual precipitation at Bushland, Texas in the southern Great Plains (weather records, USDA-ARS Conservation and Production Laboratory, Bushland, Texas). In the southern Great Plains, if the snow that falls is held in place, a larger percentage of it is usually stored as soil water than in the northern Great Plains. At southern locations, the soil seldom freezes to inhibit snowmelt water from infiltrating into the soil. At northern locations, however, much of the snowmelt may be lost as runoff because the frozen soil reduces infiltration of the water (De Jong and Steppuhn, 1983).

Snowfall in the Great Plains often is accompanied by high winds, which results in uneven distribution of snow on the landscape. The influence of topography or structures and vegetation on the relative distribution of snow cover (water equivalent) on a watershed in Canada is shown in Table 13. Stubble areas usually had a greater snow cover than fallow areas because the stubble effectively trapped the windborne snow. Bare soil areas retain little or no snow when snowfall is accompanied by wind. Consequently, soil water storage from snow is usually greater in stubble areas than on bare soil areas. Because of snow blown from bare

Table 13. Relative distribution of snow cover (water equivalent) in the Creighton watershed, Bad Lake Basin, in Canada

Land type and use	*Percentage*
Level plains	
Fallow	55
Stubble	71
Pasture	59
Hilltops	
Fallow	20
Stubble	48
Pasture	30
Gradual slopes	
Fallow	68
Stubble	69
Pasture and brush	79
Small draws	
Fallow	132
Stubble, pasture, and brush	138
Steep slopes	
Pasture and brush	253
Farmyards	
Accumulated total snowfall (Nipher shielded gage)	100
Average for entire watershed	77

Reproduced from De Jong and Stepphun, *Dryland Agriculture* pp. 89–104, 1983, by permission of the Am. Soc. Agronomy, Crop Sci. Soc. Am., and Soil Sci. Soc. Am.

soil areas and trapped in stubble areas, soil water storage from snowmelt has exceeded 100% in some stubble areas (Smika and Whitfield, 1966).

The value of surface residues for trapping snow and, consequently, increasing soil water storage is apparent from the foregoing discussion. Surface residues are most effective for trapping snow when they are standing. Other residue management practices for increasing snow trapping and conserving snowmelt include (1) alternating relatively narrow strips of tall (greater than 30 cm) stubble with areas of short (or no) stubble (De Jong and Steppuhn, 1983; Willis *et al.*, 1983); (2) establishing wind barriers of tall stubble, such as rows of corn or sunflower (De Jong and Steppuhn, 1983), sudangrass (Greb, 1983), tall wheatgrass (De Jong and Steppuhn, 1983 Greb, 1983; Willis *et al.*, 1983), and field shelterbelts (Greb, 1983; Willis *et al.*, 1983); (3) seeding winter wheat into standing spring wheat stubble (Willis *et al.*, 1983); and (4) chiseling stubble fields in the fall to below the frost zone (30 cm deep) to enhance water infiltration into nonfrozen soil at the bottom of the chisel furrow (Ramig *et al.*, 1983). Use of rows of trees or bushes are not acceptable as snow barriers in regions where water conservation is required because these plants extend roots into adjacent cropped areas and, therefore, extract soil water that could be used for crop production (Greb, 1983). Also, snow distribution with these plants is seldom uniform.

X. Summary and Conclusions

The value of water stored in soil at planting time for obtaining favorable crop yield in arid and semiarid regions has long been recognized. Early attempts to increase soil water storage involved tillage methods (clean), row spacings, fallowing, crop rotations, etc., and responses of various crops to these practices. The value of crop residues for conserving water was not realized until the late 1930s, and use of residues did not become practical under field conditions until the stubble mulch tillage system was introduced in the 1940s. Even greater emphasis on using surface residues for increasing soil water storage developed when herbicides became available for controlling weeds, thereby reducing the need for tillage to control weeds.

The additional water stored in soil due to maintaing crop residues on the surface has resulted from increased water infiltration and reduced soil water evaporation. In areas where snowfall comprises a significant part of the annual precipitation, maintaining standing residues to trap and hold snow on the land allows snowmelt water to enter the soil and is a major factor in increasing soil water storage where surface residues are present.

The additional water stored in soil where surface residues were maintained on the surface has resulted in increased crop yields at many locations. The surface residues have also provided protection against soil erosion by wind and water. Although water and soil resource conservation has resulted from maintenance of surface residues in most cases, residue production of some crops under rain fed conditions in arid and semiarid regions may not be great enough to have a major impact on soil and water conservation. In other cases, cropping systems necessary for adequate residue production are not yet available. Consequently, additional research is needed to develop suitable systems of water and soil conservation for all soils so that the crop production potential will be maintained or improved. Only by adequately conserving our soil and water resources can we be assured that future generations will have adequate resources for producing the food and fiber that they will require.

References

Allen, R.R., J.T. Musick, F.O. Wood, and D.A. Dusek. 1975. No-till seeding of irrigated sorghum double cropped after wheat. *Trans. Am. Soc. Agr. Eng.* 18:1109–1113.

Allen, R.R., J.T. Musick, and A.F. Wiese. 1976. Limited tillage of furrow irrigated winter wheat. *Trans. Am. Soc. Agr. Eng.* 19:234–236, 241.

Allen, R.R. J.T. Musick, and D.A. Dusek. 1980. Limited tillage and energy use with furrow-irrigated grain sorghum. *Trans. Am. Soc. Agr. Eng.* 23:346–350.

Baker, L.O., J.L. Krall, T.S. Aasheim, and T.P. Hartman. 1956. Chemical summer fallow in Montana. *Down to Earth* 11:21.

Bennett, O.L. 1977. Conservation tillage in the Northeast. *J. Soil Water Cons.* 32:9–12.

Blevins, R.L., D. Cook, S.H. Phillips, and R.E. Phillips. 1971. Influence of no-tillage on soil moisture. *Agron. J.* 63:593–596.

Bond, J.J., and W.O. Willis. 1969. Soil water evaporation: Surface residue rate and placement effects. *Soil Sci. Soc. Am. Proc.* 33:445–448.

Bond, J.J., and W.O. Willis. 1970. Soil water evaporation: First stage drying as influenced by surface residue and evaporation potential. *Soil Sci. Soc. Am. Proc.* 34:924–928.

Bond, J.J., and W.O. Willis. 1971. Soil water evaporation: Long-term drying as influenced by surface residue and evaporation potential. *Soil Sci. Soc. Am. Proc.* 35:984–987.

De Jong, E., and H. Steppuhn. 1983. Water conservation: Canadian prairies. In: H.E. Dregne and W.O. Willis (eds.), *Dryland agriculture. Agronomy Monograph* 23:89–104.

Duley, F.L., and Kelly, L.L. 1939. Effect of soil type, slope, and surface conditions on intake of water. Res. Bull. No. 112 Agr. Exp. Stn., Univ. of Nebraska, Lincoln.

Duley, F.L., and J.C. Russel. 1939. The use of crop residues for soil and moisture conservation. *J. Am. Soc. Agron.* 31:703–709.

Greb, B.W. 1974. Yield response to fall weed control in new wheat stubble in a fallow-wheat rotation. In: B.L. Bohmont (ed.), *Fourth annual Colorado crop Protection Inst. Proc.* Colorado State University, Fort Collins, Colorado, pp. 33–45.

Greb, B.W. 1978. Millet production with limited water. Publ. No. PR-78,Colorado State Univ. Exp. Stn. Ft. Collins, Colorado, pp. 1–3.

Greb, B.W. 1979. Reducing drought effects on croplands in the west central Great Plains. Info. Bull. No. 420, U.S. Dept. Agr., U.S. Governement Printing Office, Washington, D.C., pp. 1–31.

Greb, B.W. 1983. Water conservation: Central Great Plains. In: H.E. Dregne and W.O. Willis (eds.), *Dryland agriculture. Agronomy* 23:57–72.

Greb, B.W., D.E. Smika, and A.L. Black. 1967. Effect of straw-mulch rates on soil water storage during summer fallow in the Great Plains. *Soil Sci. Soc. Am. Proc.* 31:556–559.

Griffith, D.R., J.V. Mannering, and W.C. Moldenhauer. 1977. Conservation tillage in the eastern Corn Belt. *J. Soil Water Cons.* 32:20–28.

Hallsted, A.L., and O.R. Mathews. 1936. Soil moisture and winter wheat with suggestions on abandonment. Bull. No. 273 Kansas Agr. Exp. Stn., 44 pp.

Johnson, W.C. 1964. Some observations on the contribution of an inch of seeding time soil moisture to wheat yield in the Great Plains. *Agron. J.* 56:29–35.

Johnson, W.C., and R.G. Davis. 1972. Research on stubble-mulch farming of winter wheat. USDA-ARS Conserv. Res. Rep. No. 16, U.S. Government Printing Office, Washington, D.C., pp 1–21.

Johnson, W.C., C.E. Van Doren, and E. Burnett. 1974. Summer fallow in the southern Great Plains. *In: Summer fallow in the western United States.* Cons. Res. Rept. No. 17, USDA Agr. Res. Serv., Washington, D.C. p. 97.

Jones, J.N., Jr., J.E. Moody, and J.H. Lillard. 1969. Effects of tillage, no-tillage, and mulch on soil water and plant growth. *Agron. J.* 60:719–721.

Mannering, J.V., and L.D. Meyer. 1963. The effects of various rates of surface mulch on infiltration and erosion. *Soil Sci. Soc. Am. Proc.* 27:84–86.

Mannering, J.V., D.R. Griffith, and C.B. Richey. 1975. Tillage for moisture conservation. Paper No. 75-2523, Am. Soc. Agr. Eng., St. Joseph, Michigan.

McCalla, T.M., and J.T. Army. 1961. Stubble mulch farming. *Adv. Agron.* 13;125–196.

Musick, J.T., A.F. Wiese, and R.R. Allen. 1977. Management of bed-furrow irrigated soil with limited- and no-tillage systems. *Trans. Am. Soc. Agr. Eng.* 20:666–672.

Nelson, L.R., R.N. Gallaher, and R.R. Bruce. 1976. Corn forage yields in double-cropping systems. *Fertilizer Solutions* 20:56–63.

Phillips, R.E. 1974. Soil water, evapotranspiration and soil temperature in no-tilled soil. In: R.L. Blivens and G.L. Thomas (eds.), *Proceedings, no-tillage research conference.* Univ. of Kentucky, Lexington, pp. 6–14.

Phillips, W.M. 1964. A new technique of controlling weeds in sorghum in a wheat-sorghum-fallow rotation in the Great Plains. *Weeds* 12:42–44.

Phillips, W.M. 1969. Dryland sorghum production and weed control with minimum tillage. *Weed Sci.* 17:451–454.

Ramig, R.E., R.R. Allmaras, and R.I. Papendick. 1983. Water conservation: Pacific Northwest. In: H.E. Dregne and W.O. Willis (eds.), *Dryland Agriculture. Agronomy Monograph* 23:105–124.

Reicosky, D.C., D.K. Cassel, R.L. Blevins, W.R. Gill, and G.C. Naderman. 1977. Conservation tillage in the Southeast. *J. Soil Water Cons.* 32:13–19.

Russel, J.C. 1939. The effect of surface cover on soil moisture losses by evaporation. *Soil Sci. Soc. Am. Proc. 4:65–70.*

Smika, D.E. 1976a. Mechanical tillage for conservation fallow in the semiarid Central Great Plains. In: B.W. Greb (ed.), Conservation tillage, Proc. Great Plains Workshop. Publ. No. 77, Great Plains Agr. Council, Fort Collins Colorado, pp. 78–91.

Smika, D.E. 1976b. Seed zone soil water conditions with reduced tillage in the semiarid Central Great Plains. Proc. 7th Conf. Int. Soil Tillage Res. Org. 37.1-37.6, Uppsala, Sweden. June 13–18, 1976.

Smika, D.E. 1983. Soil water change as related to position of wheat straw mulch on the soil surface. *Soil Sci. Soc. Am. J.* 47:988–991.

Smika, D.E., and C.J. Whitfield. 1966. Effect of standing wheat stubble on storage of winter precipitation. *J. Soil Water Cons.* 21:138–141.

Smika, D.E., and G.A. Wicks. 1968. Soil water storage during fallow in the Central Great Plains as influenced by tillage and herbicide treatments. *Soil Sci. Soc. Am. Proc.* 32:591–595.

Unger, P.W. 1976. Surface residue, water application, and soil texture effects on water accumulation. *Soil Sci. Soc. Am. J.* 40:298–300.

Unger, P.W. 1978. Straw-mulch rate effect on soil water storage and sorghum yield. *Soil Sci. Soc. Am. J.* 42:486–491.

Unger, P.W. 1984. Tillage and residue effects on wheat, sorghum, and sunflower grown in rotation. *Soil Sci. Soc. Am. J.* 48:885–891.

Unger, P.W., and J.J. Parker. 1975. No-tillage dryland grain sorghum after irrigated wheat with intervening fallow. Prog. Rep. No. 3330, Texas Agr. Exp. Stn., College Station, Texas, pp. 43–54.

Unger, P.W., and J.J. Parker. 1976. Evaporation reduction from soil with wheat, sorghum, and cotton residues. *Soil Sci. Soc. Am. J.* 40:938–942.

Unger, P.W., and A.F. Wiese. 1979. Managing irrigated winter wheat residues for water storage and subsequent dryland grain sorghum production. *Soil Sci. Soc. Am. J.* 43:582–588.

Unger, P.W., R.R. Allen, and A.F. Wiese. 1971. Tillage and herbicides for surface residue maintenance, weed control, and water conservation. *J. Soil Water Cons.* 26:147–150.

Van Doren, D.M., Jr., and R.R. Allmaras. 1978. Effect of residue management practices on the soil physical environment, microclimate, and plant growth. In: W.R. Oschwald (ed.), *Crop residue management systems.* Spec. Publ. No. 31, Am. Soc. Agron, Madison, Wisconsin, pp. 49–83.

Van Doren, D.M., Jr., G.B. Triplett, Jr., and J.E. Henry. 1976. Influence of long term tillage, crop rotations, and soil type combinations on corn yield. *Soil Sci. Soc. Am. J.* 40:100–105.

Wicks, G.A., and D.E. Smika. 1973. Chemical fallow in a winter wheat-fallow rotation. *J. Weed Sci. Soc. Am.* 21:97–102.

Wiese, A.F., and D.W. Staniforth. 1973. Weed control in conservation tillage. In: A.B. Linford (ed.), *Conservation tillage, the Proceedings of a National Conference.* Soil Cons. Soc. Am., Ankeny, Iowa, pp. 108–114.

Willis, W.O., A.Bauer, and A.L. Black. 1983. Water conservation: Northern Great Plains. In: H.E. Dregne and W.O. Willis (eds.), *Dryland Agriculture. Agronomy Monograph* 23:73–88.

Physical Properties and Processes of Puddled Rice Soils[1]

Pradeep K. Sharma and Surajit K. De Datta*

I. Introduction

In most Asian countries, wetland tillage for rice (*Oryza sativa* L.) puddling has become almost synonymous with rice culture (Sanchez, 1976), although in other parts of the world—the United States, Australia, parts of Europe, and some Asian countries—rice land is prepared dry and flooded later (Savant and De Datta, 1982). Puddling, in general, refers to the destruction of soil aggregates into ultimate soil particles at a moisture content near saturation. Buehrer and Rose (1943) defined puddling as "the destruction of the aggregated condition of the soil by mechanical manipulation within a narrow range of moisture contents above and

[1]Contribution from the Department of Agronomy, The International Rice Research Institute, Los Baños, Laguna (Mail address: P.O. Box 933, Manila), Philippines.

*Former Postdoctoral Fellow [present address: Regional Research Station, Dhaulakuan-173 001, Dist. Sirmour (H.P.), India], and Principal Scientist and Head, Department of Agronomy, International Rice Research Institute.

below field capacity (0.03 MPa), so that soil aggregates lose their identity and the soil is converted into a structurally more or less homogeneous mass of utlimate particles." According to Bodman and Rubin (1948) "puddling is the mechanical reduction in the apparent specific volume of soil." From the farmer's point of view, however, puddling is simply the mixing of saturated soil with flood water to make it soft for transplanting and impervious to water (De Datta, 1981). Puddling, although a capital- and energy-intensive process, is practiced to achieve certain objectives such as weed control, ease of transplanting, and reduction in percolation losses of water and nutrients. The advantages and disadvantages of puddling in rice-based cropping systems are summarized in Table 1.

II. The Puddling Process

During puddling, soils are subjected to two kinds of deforming stresses: (1) the normal stress (load) associated with compression and (2) tangential stress causing shear (Bodman and Rubin, 1948; Baver, 1956). The compression is most effective below the upper plastic limit (moisture content at which the soil–water system can flow as a sticky fluid paste); shearing effects dominate above the upper plastic limit. The work done (*W*) in puddling can be represented by the following equation (Bodman and Rubin, 1948):

$$W = -\int_{Vi(c)}^{Vf(c)} P\,dv \quad - \int_{Vf(c)}^{Vz} P\,dv \quad + \frac{1}{M_s}\int_{\theta_0}^{\theta_z} T\,d\theta \qquad [1]$$

total work done in puddling	=	work done by normal stress before shear	+	work done by normal stress during shear	+	work done by tangential stress during shear

where *P* is the normal stress, M_s is the mass of dry *soil*, *T* is the torque, θ is the angular rotation of a piston, *v* is the apparent specific volume, *Vi*(*c*) is the initial specific volume at the beginning of normal stress, *Vf*(*c*) is the final specific volume due to normal stress, *Vz* is the terminal value due to shear, θ_0 is the zero angular displacement of a piston, and θ_z is the final displacement of a piston.

The ease and degree of puddling depend on soil wetness, soil type, nature of the implement used, and cultural practices. Maximum puddling occurs at moisture contents between field capacity and saturation (McGeorge, 1937; Bodman and Rubin, 1948; Koenigs, 1963). At this moisture (1) cohesion *within* soil aggregates is minimal, which helps increase the number of shear planes in soil. Wetting of a dry soil causes an uneven swelling and explosion of aggregates (Cernuda *et al.*, 1954), resulting in increased number of shear planes, and (2) the cohesion *between* the aggregates (and clods) is maximum, which restricts their

Table 1. Advantages and disadvantages of puddling in rice-based cropping systems (adapted from De Datta and Barker, 1978; De Datta, 1981)

A. Advantages
1. Improves weed control by primary and secondary tillage through puddling action
2. Eases transplanting
3. Establishes a reduced soil condition, which improves soil fertility and fertilizer management
4. Reduces draft requirements for primary and secondary tillage
5. Reduces percolation losses, conserving water from rainfall and irrigation
6. Takes advantage of monsoon rains by the time puddling has been completed

B. Disadvantages
1. Requires a large volume of water
2. Hinders regeneration of soil structure and insertion of an upland crop following lowland rice
3. Impedes root development

movement along each other and along the implement. Consequently, maximum energy is transferred for the destruction of aggregates.

Soils with high cohesion within aggregates, because of the presence of such stabilizing agents as hydrous oxides of Fe and Al, calcium carbonates, and organic matter, involve considerably more mechanical work to become well puddled (Buehrer and Rose, 1943; Koenigs, 1963; Rengasamy *et al.*, 1974). High clay content favors puddling (Buehrer and Rose, 1943; Allman and Kohnke, 1947; Bodman and Rubin, 1948; Koenigs, 1961; Zein El Abedine *et al.*, 1969). Even small amounts of silt and clay are enough to achieve puddling in sandy soils (Koenigs, 1961). Smectite clays, because of their swelling properties, are easier to puddle than kaolinitic clays (Buehrer and Rose, 1943; Harris *et al.*, 1966). Sodium-saturated clays are easier to puddle than Ca-saturated clays (McGeorge, 1937).

The degree of puddling further depends on the type of the implement used and the intensity of puddling (Taneja and Patnaik, 1962; Kokubun *et al.*, 1969). Rotary implements are expected to give better puddling than are plows (Gliemeroth, 1953, cited by Koenigs, 1963), because rotary motion produces continuous change in the direction of shear planes, which helps in finding the weakest fracture planes in the clods. Sinha (1964) reported greater puddling in a clay soil with a rotary puddler than with a plow. According to Tyagi *et al.* (1975), a power tiller with rotovator gave better puddling in a loamy sand soil than a moldboard or a country plow. In contrast to this, however, Agarwal *et al.* (1978), in a sandy loam soil, reported that a country plow followed by planking was better than a rotary puddler. Prihar *et al.* (1976) obtained no difference in puddling of a sandy loam soil with a country plow, disk harrow, angular puddler, and a rotovator. These diverse conclusions may be attributed to the differences in soil types and puddling indices used.

III. Puddling Indices

No systematic work has been reported on the evaluation of degree of puddling and the puddlability of soils. Indian farmers have used a variety of qualitative methods to determine the degree of puddling, including (Sinha, 1964): (1) when a pitcher is thrown on a puddled soil, it should not break; (2) a man walking over a well-puddled soil should not feel any roughness under his feet.

Bodman and Rubin (1948) gave a quantitative dimension to puddling. They introduced the concept of "puddlability" which refers to the susceptibility of soil to puddling. Puddlability may be defined as the change in the apparent specific volume of soil per unit of energy expended; the decrease in the apparent specific volume is used as a measure of degree of puddling. Taneja and Patnaik (1962) studied the shrinkage in volume of puddled soil, subjected to centrifugation at 2000 rpm, to characterize the degree of puddling. Bhole (1963) used viscosity of puddled soil as an index of degree of puddling. Sinha (1964) developed an "index of puddling" based on the dispersion of soil particles. The "index of puddling" refers to the ratio of volume of soil suspension (puddled soil) after and before settling (for about 48 hr). A higher index value suggests greater puddling.

Naphade and Ghildyal (1971) used aggregate size distribution of puddled soil, as determined by Bouyoucos hydrometer, to evaluate the degree of puddling. Decrease in saturated hydraulic conductivity and percolation rates on puddling have also been used as indices of the degree of puddling (Rane and Varade, 1972; Tyagi *et al.*, 1975; Agarwal *et al.*, 1978).

However, none of these indices is widely used because of limitations. For example, puddling may not always decrease the apparent specific volume of soil, as suggested by Bodman and Rubin (1948). Puddling decreases or increases the apparent specific volume of soil depending on the nature of the soil colloids, ionic concentrations of the soil solution, and aggregation status of the soil before puddling (Greenland, 1981; Sharma and De Datta, 1985). Soil dispersion, among other factors, is also influenced strongly by the electrolyte concentration of the soil solution. Further, if the initial dispersion of soil on puddling is low, the drop in its volume on settling will also be low, and the "index of puddling" will indicate a high degree of puddling, which otherwise is not the case. Also, this soil might not have the advantage of reduced percolation losses. In situ measurements of decline in hydraulic conductivity or percolation rates as an index of degree of puddling may not be useful because the same effect can be achieved through soil compaction instead of making it thoroughly pulverized (Ghildyal and Satyanarayana, 1965). Hence, there is still a need to develop a suitable puddling index that combines the intents of puddling, such as softness of soil (for ease of transplanting) and reduced percolation rate (for water and nutrient economy). A combined evaluation of bulk density and percolation rate may be appropriate. Change in the ratio of silt plus clay dispersed in water to the actual silt plus clay in soil, before and after puddling, may also be tested as an index of puddling.

IV. Effects of Puddling

Puddling influences the physical, chemical, and microbiological properties of soils. Extensive reviews are available on chemical and electrochemical (Ponnamperuma, 1972, 1978; Patrick and Reddy, 1978; Tadano and Yoshida, 1978; De Datta, 1981; Savant and De Datta, 1982) and microbiological properties (Sethunathan and Siddharamappa, 1978; Watanabe, 1978; T. Yoshida, 1978) of puddled rice soils. In this contribution we review and discuss physical processes involved in puddling and the resulting changes in soil physical properties in relation to rice growth. The puddling effects are divided into short term and long term.

A. Short-Term Effects

1. Soil Structure

The most obvious effect of puddling is the destruction of soil aggregates and peds into plastic mud (McGeorge, 1937; Chang, 1941; Robinson and Page, 1951; Beacher and Strickling, 1955; Swanson *et al.*, 1955; Ghildyal, 1982). Wetting of initially dry soil causes uneven swelling in aggregates and explosion caused by entrapped air. The result is the slaking of aggregates. In addition, depending on soil type, flooding may weaken interparticle bonding by increasing the solubility of iron and manganese oxides and decomposition of organic matter (Kawaguchi *et al.*, 1956; Kawaguchi and Kita, 1957), thereby dispersing soil aggregates; although in some cases granulated structure was maintained even after several months of flooding (Sanchez, 1968). Repeated plowings and harrowings or use of puddlers partially or completely destroys soil aggregates, depending on their stability. A well-aggregated porous soil is converted into a plastic mud of massive structure.

Chaudhary and Ghildyal (1969) reported a drop in mean weight diameter of aggregates from 1.70 mm to 0.36 mm due to puddling. In a laboratory study, about 40% of the aggregates were completely broken down to fractions less than 0.05 mm due to puddling; and all aggregates were smaller than coarse sand (Naphade and Ghildyal, 1971). According to Yun-sheng (1983), changing from double cropping rice and wheat (*Triticum aestivum* L.) to triple cropping rice, rice, and wheat lowered the amount of microaggregates by 50% between 1965 and 1976.

The puddled layer is neither structurally nor chemically uniform, but information on stratification is not adequate. According to Saito and Kawaguchi (1971a), the 0- to 15-mm puddled layer is composed of fine particles, the middle thin and porous layer is of sandy shingles, and the massive lower layer lacks particle differentiation. In a puddled layer, clay particles, or clusters thereof, are oriented in parallel rows and are surrounded by water-saturated capillary pores. Thus, soil body is reduced to a two-phase system, solid and liquid; the gaseous phase is either eliminated or entrapped in storage and residual pores (Buehrer and Rose, 1943; Saito and Kawaguchi, 1971b).

Some reports suggest that soil structure lost during puddling is not only regenerated but is further improved during crop growth, under submerged conditions (Chaudhary and Ghildyal, 1969). However, most of such studies have determined stability of aggregates formed by breaking the dried soil puddle. Aggregate stability is significantly affected by drying (Swanson *et al.*, 1955; Baver, 1956; Quirk, 1978). Stability may increase on drying due to the precipitation of previously reduced Fe and Mn oxides and hydroxides. Hence, these results may be used with caution.

Destruction of aggregation during puddling triggers a series of changes in physical properties of soils that affect not only plant growth but also in some cases tractor use. In continuously submerged clay soils, progressive deterioration of soil structure by puddling has made soils excessively soft, creating problems in tractor use (Kisu, 1978).

2. Bulk Density and Soil Strength

The change in bulk density caused by puddling depends on the aggregation status of the soil before puddling. If a parallel-oriented, closely packed structure is produced from a well-aggregated open structure, bulk density increases and vice versa (Aylmore and Quirk, 1962). In general, strong interparticle forces favor well-oriented structure, whereas weak interparticle forces favor an open gel structure (Greenland, 1981). Puddling of a well-aggregated porous soil results in a massive structure of increased bulk density (Bodman and Rubin, 1948; Dutt, 1948; Naphade and Ghildyal, 1971; Ghildyal, 1978, 1982). Change in the soil bulk density depends not only on the intensity of puddling but also on the cropping system and crop residue management (Fagi and De Datta, 1983). In swelling clays, even submergence, without puddling, substantially decreases bulk density (Virmani *et al.*, 1982). Bulk density increases on drying because of soil shrinkage. A dried puddled soil is very compact and hard and develops broad and deep cracks, depending on the nature and amount of clays. The cracks usually form a hexagonal pattern on the soil surface; their spacing depends on the ability of soil to deform under tensile stresses, which are generated parallel to the soil surface, when the soil dries and shrinks (Lachenbruch, 1961). Puddling of such a soil brings the soil particles back in suspension and lowers the bulk density. In a field experiment, immediately after puddling, the bulk density of a lowland clay soil was reduced from 0.83 to 0.53 mg m^{-3}, and from 1.16 to 0.81 mg m^{-3} in a clay loam soil (Table 2). Saito and Kawaguchi (1971c) observed similar results. However, bulk density may increase with time even when soil is submerged, because of the settling of particles (Table 2). Soil particles may settle and consolidate because of reflocculation of dispersed clay, depending possibly on soil texture, type of clay, and the ionic concentrations of soil solution (zeta potential). The main flocculating and cementing agents in puddled soils are active Fe, Mn, and Al and organic complexes (Kawaguchi and Kita, 1957; Saito, 1985). Settling is faster in sandy soils or in soils with kaolinitic mineralogy (Moormann and van Breemen, 1978). A report from Surinam (Scheltema, 1974) suggests an increase in bulk density of puddled soil with time, with a consequent decrease in soil water content from 90 to 100%, under continuously flooded conditions, to 20 to 60%.

Soil strength is a measure of resistance that must be overcome to cause a deformation in soil. The resistance is to both volumetric compression and linear deformation and depends on moisture content, texture, type and amount of clay, arrangement of particles in the soil matrix, and ionic species and concentrations. Soil strength decreases with increase in moisture content and increases with bulk density (Kumar *et al.*, 1971; Kisu, 1978). Soil strength is rapidly measured in situ with cone penetrometers. However, penetrometer readings should be interpreted very carefully, because sometimes a wide combination of soil properties may give the same values of cone index (Dexter and Woodhead, 1985).

Soil strength in the puddled layer may be near zero because of high moisture content and loosely arranged soil particles. Figure 1 shows soil strength profiles in a puddled and nonpuddled soil at the time of rice transplanting. In the top 10-cm puddling layer, soil strength was 0.1 MPa as against 0.5 MPa in the nonpuddled soil. Between 10 and 15 cm, soil strength in the nonpuddled layer was twice as high as in the puddled layer.

Bulk density and soil strength are negatively correlated with growth and grain yield of transplanted rice (R.K. Gupta and Jaggi, 1979; Huang, 1982). Figure 2 shows relationships between rice grain yield, root length

Table 2. Effect of puddling on bulk density of soils

	Bulk density ($Mg\ m^{-3}$)											
	M17a						*UE2*					
	1 DAT[a]			*60 DAT*[a]			*1 DAT*[a]			*60 DAT*[a]		
Depth (cm)	*P*[b]	*NP*[c]	*LSD*[d]	*P*[b]	*NP*[c]	*LSD*[d]	*P*[b]	*NP*[c]	*LSD*[d]	*P*[b]	*NP*[c]	*LSD*[d]
0 to 10	0.53	0.83	0.14[e]	0.67	0.89	0.10[e]	0.81	1.16	0.19[e]	1.18	1.21	ns
10 to 20	0.68	0.91	0.17[f]	0.99	1.02	ns	1.09	1.23	0.09[f]	1.34	1.39	ns
20 to 30	1.02	1.00	ns	1.04	1.04	ns	1.27	1.29	ns	1.33	1.38	ns

[a]DAT = days after transplanting.
[b]P = puddled.
[c]NP = nonpuddled.
[d]LSD = least significant difference.
[e] = significant at 1% level.
[f] = significant at 5% level.
Reproduced from Sharma and De Datta, *Soil Science of America Journal*, Vol. 49, pp. 1451–1457, 1985, by permission of the Soil Sci. Soc. Am.

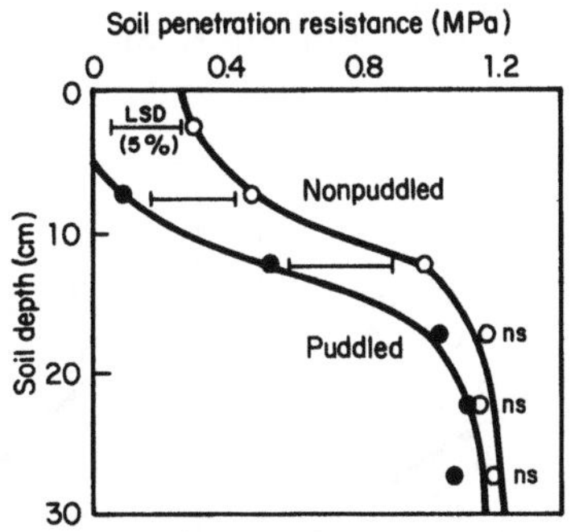

Figure 1. Effect of puddling on the penetration resistance of a clay loam soil (adapted from Sharma and De Datta, 1985). Reproduced from *Soil Science Society of America Journal,* Vol. 49, pp. 1451–1457, 1985, by permission of Soil Sci. Soc. Am.

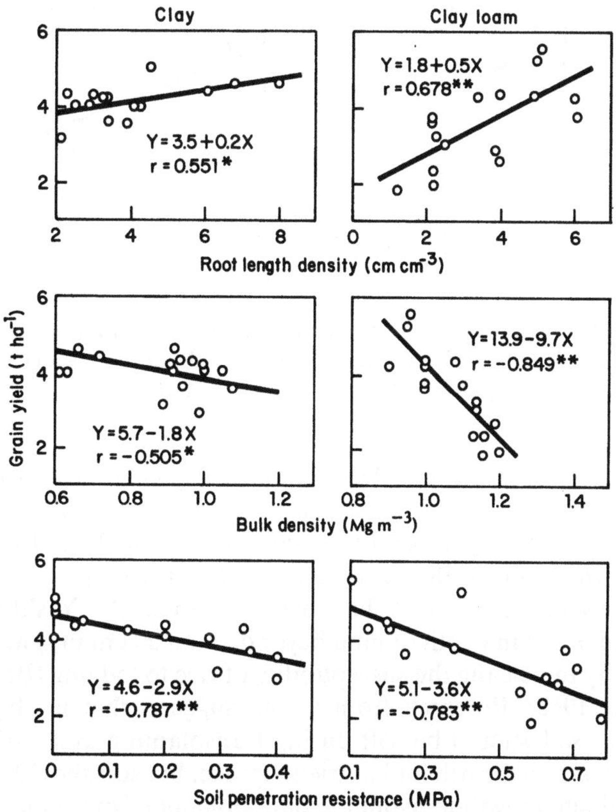

Figure 2. Relationship among root length density at harvest, bulk density, and soil strength at transplanting in the 0- to 10-cm soil layer and rice grain yield (P.K. Sharma and S.K. De Datta, unpublished). * = significant at 5% level; **significant at 1% level.

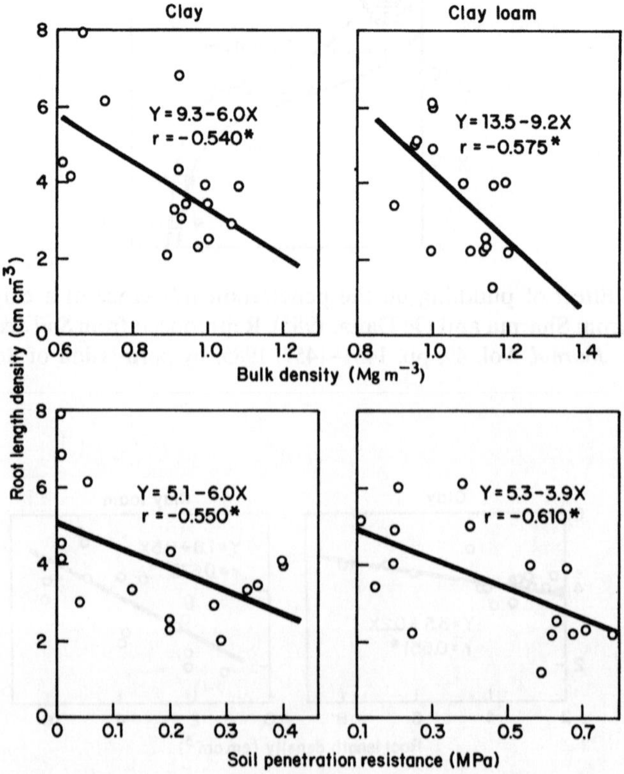

Figure 3. Relationship between bulk density and soil strength at transplanting, and root length density of rice at harvest in the 0- to 10-cm soil layer (P.K. Sharma and S.K. De Datta, unpublished). * = significant at 5% level.

density at harvest, and bulk density and soil penetration resistance at transplanting, in the top 10-cm soil layer. With greater bulk density and soil strength, root length density is reduced (Figure 3) and grain yield drops. A significant positive correlation between root growth and total biomass production of rice has been reported (S. Yoshida, 1981). However, increase in depth of mud beyond about 30 cm may again lower rice yields by increasing the susceptibility of rice to lodging (Briones and Raymundo, 1962). Research from China suggests that for higher rice grain yields, soil should be soft during transplanting and early growth stages to favor root growth and nutrient uptake; but soil should be firm in middle and late growth to avoid the development of unproductive tillers and to support the plants (Xian-Ling, 1984, personal communication).

3. Soil Porosity

Change in porosity by puddling depends on the change in the orientation of soil particles in puddled layer. As discussed earlier, if a parallel-oriented structure is produced on puddling from an initially open-gel structure, total porosity will decrease, and vice versa. Total pore volume usually increases from puddling, perhaps temporarily. For example, at the International Rice Research Institute (IRRI) farm, puddling increased total porosity from 69 to 80% in clay, and from 56 to 69% in clay loam soil. However, Chang (1941) reported a decline in total porosity from 48 to 26%.

The accurate measurement of pore-size distributions in puddled soils is difficult. The capillary rise equation, normally used in rigid porous systems, is difficult to apply in a nonrigid, swollen system, where structure changes as water is desorbed, i.e., the soil shrinks. Other methods used in the study of pore-size distribution have their own merits and demerits (Greenland, 1981). Consequently, very few direct studies on pore-size distribution in puddled soils have been reported. Most data come from laboratory studies. Bodman and Rubin (1948) reported 90 to 100% reduction in the noncapillary porosity in a silty clay loam soil. Jamison (1953) observed about 80% reduction in pores >10 μm in a clay soil, with a corresponding increase in the water-filled pores. Aomine and Shiga (1959) studied several Japanese rice soils after crop harvest and identified a massive layer A_{p1} with very few pores in the top 1 to 2 cm, followed by a relatively more porous A_{p2} about 10-cm deep, underlain again by a massive A_{p3}. Figure 4 compares pore-size distribution in puddled and aggregated soils at IRRI farm. For puddled samples, soil cores were collected after the harvest of lowland rice; for aggregated soil samples, cores were collected after the harvest of maize. In both soils, puddling decreased pores > 30 μm (transmission pores) by about 83% and increased pores between 0.6 and 30 μm (storage pores) and < 0.6 μm (residual pores) by 7% and 52%, respectively, as compared to aggregated soil. Data of Warkentin (1962) also suggest increase in storage pores of soils on puddling.

Pore-size distributions and rigidity of the pore system are as important for growth and development of lowland rice roots as for upland crops. Research in India demonstrated that under submerged conditions rice roots penetrated extensively in soils with large total porosity and a high proportion of pores > 150 μm diameter (Kar and Varade, 1972; Kar and Ghildyal, 1975; Kar *et al.*, 1979a). The influence of size and rigidity of pores on root growth became less evident with finer soil particles and increased porosity (Kar *et al.*, 1979a). Soil compaction, on the other hand, reduced root growth of rice (Kar *et al.*, 1976). A strong interaction of bulk density, soil moisture, and soil temperature in relation to number, volume, and depth of penetration of rice roots was observed (Kar *et al.*, 1976; Kar *et al.*, 1979b). Puddling of soil, therefore, improves root growth of rice by decreasing the rigidity of pores and soil resistance to penetration.

Changes in pore-size distribution due to puddling strongly influence other soil physical processes, such as gaseous exchange, water retention and transmission, and evaporation losses from soils.

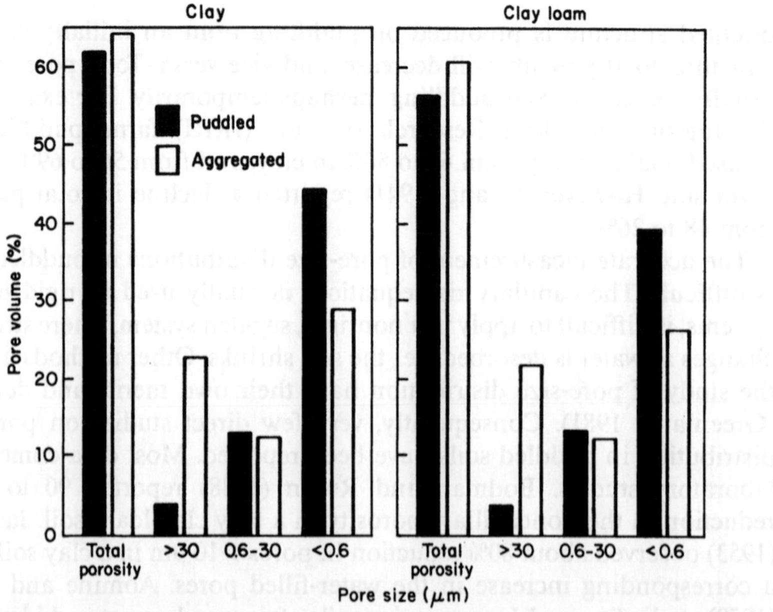

Figure 4. Effect of puddling on pore-size distributions of soil (P.K. Sharma and S.K. De Datta, unpublished).

4. Gaseous Exchange

Two main processes of gaseous exchange between soil and outer atmosphere are recognized: mass flow and diffusion. The contribution of mass flow, where driving force is the total pressure gradient of gases, is almost negligible as compared to diffusion, which depends on the partial pressure of gases. The diffusion of gases is described by Fick's law:

$$q_i = -aD_{ai}(T_o, P_o)\left(\frac{T}{T_o}\right)^m\left(\frac{P_0}{P}\right)\frac{dC_i}{dz} \qquad [2]$$

where q_i is the diffusion rate of gas *i*, *a* is a porosity factor, D_a is the diffusion coefficient in air at absolute temperature T_o and pressure P_o, *T* is the actual absolute temperature, *P* is the actual barometric pressure, C_i is the concentration of gas *i* in soil air, *z* is depth, and $m = 1.75$ for O_2 and 2.00 for CO_2 (for diffusion in air). Diffusion coefficients of gases in water are about 10^{-4} of their values in air. In the soil–water system the diffusion coefficient is further reduced by the presence of solid particles and by the tortuosity factor. The effective diffusion coefficient in soil water (D_{sw}) is related to the diffusion coefficient in pure water (D_w) by the following equation:

$$D_{sw} = D_w \tau \lambda V_w \qquad [3]$$

where τ is the tortuosity factor, λ is a factor accounting for the interaction between gas and soil (usually = 1), and V_w is the moisture content.

Table 3. Diffusion constants of CO_2, O_2, and N_2 in air and water and solubilities of these compounds in water at 20°C

	Diffusion constant ($m^2\ s^{-1}$)		
Gas	*Air*	*Water*	*Solubility (absorption coefficient)*
CO_2	16.1×10^2	0.177	0.878
O_2	20.5×10^2	0.180	0.031
N_2	20.5×10^2	0.164	0.015

Reprinted with permission from De Jong and Paul, in R.W. Fairbridge and C.W. Finkl, Jr. (eds.), *The Encyclopedia of Soil Science, Part I*, pp 10–21. Copyright 1979 by Dowden, Hutchinson & Ross, Inc.

In a water-saturated soil, the driving force for the diffusion is the concentration gradient across the gas–liquid interface. The concentration of gas in the liquid phase is related to the concentrations in gaseous phase through the solubility constants (Table 3). Consequently, the same concentration difference between two gas phases separated by a water film will move about 16 times as much CO_2 as O_2.

Thus, in a submerged puddled soil, the exchange of gases, especially of O_2, between soil and the outer atmosphere is severely restricted. The net result is a decline in O_2 concentration and a buildup of CO_2 concentration in soil. Some soils may generate as high as 2.5 tons of CO_2 per hectare within 3 weeks of submergence (Ghildyal, 1982). In puddled soils, rice usually does not suffer from O_2 deficiency because most of the O_2 demand of the rice plant can be met through O_2 transported from aerial parts to the root through intercellular gaseous spaces (Luxmoore *et al.*, 1970). For most upland crops, the critical oxygen diffusion rate (ODR) limit is 18×10^{-5} kg $m^{-2}hr^{-1}$ (Letey and Stolzy, 1964), whereas rice can grow well even at 2.4×10^{-5} kg $m^{-2}hr^{-1}$ ODR (Kumar *et al.*, 1971). However, complete elimination of O_2 in the root medium even for 4 hr can permanently stop rice root growth (Webb and Armstrong, 1983).

5. Water Retention and Transmission

At about −0.01 MPa water potential, water retention was not much affected by puddling despite the elimination of noncapillary pores (Elrick and Tanner, 1955; R.K. Gupta and Jaggi, 1979). At lower water potentials (more negative), water retention in puddled soils always exceeds that in nonpuddled soils depending on soil texture and initial aggregation (Bodman and Day, 1943; Buehrer and Rose, 1943; Campbell, 1952; Jamison, 1953; Croney and Coleman, 1954; Warkentin, 1962; Taylor *et al.*, 1967; Taylor, 1972; Yun-sheng, 1983). Figure 5 shows differences in moisture desorption curves for puddled and dry-tilled lowland clay loam

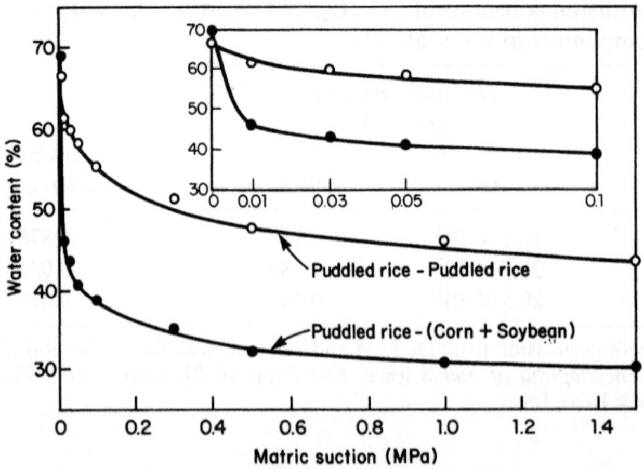

Figure 5. Effect of puddling on the soil moisture desorption curve (P.K. Sharma and S.K. De Datta, unpublished).

soil at an IRRI farm. Consequently, the available water capacity ($\theta_{0.01\ MPa} - \theta_{1.5\ MPa}$) of puddled soils is usually higher than nonpuddled soils (Jamison, 1953; Raymundo *et al.*, 1962). Drying brings soil moisture characteristics of puddled soils close to those of nonpuddled ones (Figure 6) (Taylor, 1972); moreover, resaturation does not restore to a dried puddled soil its original high water-retention capacity (Croney and Coleman, 1954). This implies that in rain-fed conditions a rice crop may

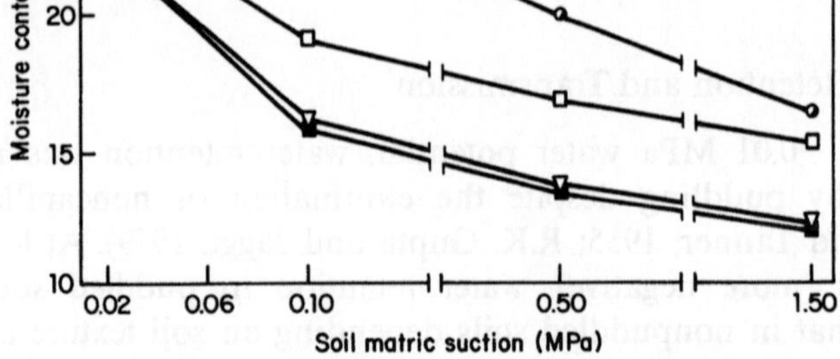

Figure 6. Effect of different predrying treatments on water characteristic curves of a silty clay loam soil. Reproduced from Taylor, *Soil Science of America Proceedings*, Vol. 36, pp. 972–973, 1972, by permission of the Soil Sci. Soc. Am.

experience more drought stress during prolonged rainless periods in puddled than in aggregated soil.

Evaporation losses from puddled soils are low because of their increased microporosity (R.K. Gupta and Jaggi, 1979; R.P. Gupta and Nagarajarao, 1982). In puddled soils, because of high water-retention capacity, more energy is needed to evaporate the same quantity of water than in upland soils. Higher unsaturated hydraulic conductivity of puddled soils (because of large microporosity) may also help in keeping the surface soil moist longer by transporting more profile water upward to the soil surface. Consequently, it may take weeks to months to dry a puddled soil as compared to a well-structured soil (Figure 7). Rice grown on a puddled soil may therefore be less affected by mild drought than rice on a granulated soil (De Datta and Kerim, 1974). Obermueller and Mikkelsen (1974) obtained higher growth and grain yield of rice on a puddled soil than on a nonpuddled aggregated soil. In a recent field experiment on a sandy loam soil, we obtained 40% higher grain yield in rain-fed transplanted rice (puddled soil) than in rain-fed dry-seeded rice (nonpuddled soil). Under a severe drought situation, however, when the soil starts shrinking and cracking, rice performs more poorly on a puddled than on a well-aggregated soil due to restricted root development (Sanchez, 1973a). However, the boundaries between *mild* and *severe* moisture stresses, which can decide the superiority of one method of land preparation over the other, are not known. Research is needed to evaluate puddled vs. nonpuddled rice under conditions of moisture stress of varying degrees and at different rice growth stages.

Close packing of soil particles in parallel orientation in puddled soils reduces their saturated hydraulic conductivity and percolation losses in rice fields (Varade and Ghildyal, 1967; Naphade and Ghildyal, 1971; De Datta *et al.*, 1973; Sanchez, 1973b; De Datta and Kerim, 1974; Asthana

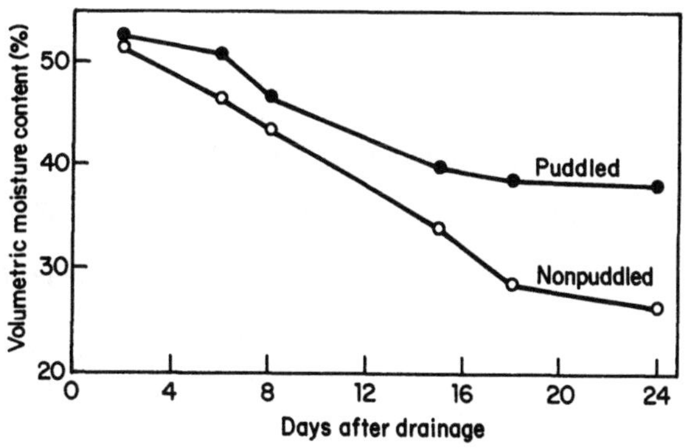

Figure 7. Drying characteristics of the top 5 cm of puddled and nonpuddled soil at rice harvest. Reprinted with permission from R.K., Gupta and Jaggi, *J. Agron. Crop Sci.,* Vol. 148, pp. 329–336. Copyright 1979 by Parey-Verlag.

Table 4. Effect of puddling on the saturated hydraulic conductivity of soils

Soil texture	*Saturated hydraulic conductivity* ($\mu m\ s^{-1}$) *Before puddling*	*After puddling*	*Source*
Sandy loam	0.53	0.06	Varade and Ghildyal (1967)
Sandy clay loam	0.53	0.05	Naphade and Ghildyal (1971)
Loam	7.08	0.44	Pande (1975)
Clay loam	0.18	0.04	Sharma and De Datta (1985)
Clay	0.03	0.02	Sharma and De Datta (1985)

and Vamadevan, 1977; Wickham and Singh, 1978; R.K. Gupta and Jaggi, 1979; Reddy and Sharma, 1979; De Datta, 1981; Hundal and De Datta, 1984a; Sharma and De Datta, 1985). Development of subsurface hardpan (Grant, 1965) or the presence of constant shallow water tables in rice fields further restricts the downward movement of water (Wickham and Singh, 1978). The effect of puddling on the water-transmission properties of soil varies with soil type, intensity of puddling, and type of implement used (Rane and Varade, 1972; Sinha *et al.*, 1981). Yun-sheng (1983) observed a decline in percolation rate from 0.10 to 0.17 $\mu m\ s^{-1}$ to 0.02 to 0.12 $\mu m\ s^{-1}$ due to shift in cropping systems from rice and wheat to rice, rice, and wheat. Some data on hydraulic conductivity and percolation rates of puddled and nonpuddled soils are given in Tables 4 and 5.

Reduction in percolation improves water and nutrient efficiency in rice fields. In puddled soils, two factors are responsible for maximizing

Table 5. Effect of puddling on percolation rates of soils

Soil texture	*Percolation rates* ($\mu m\ s^{-1}$) *Before puddling*	*After puddling*	*Source*
Clay	25	0.02	Sanchez (1973b)
Clay	0.07	0.02	IRRI (1973)
Clay	0.29	0.14	R.K. Gupta and Jaggi (1979)
Clay	0.03	0.02	Sharma and De Datta (1985)
Clay loam	0.10	0.03	Sharma and De Datta (1985)

nutrient efficiency: increased solubility (Ponnamperuma, 1972; Patrick and Reddy, 1978), and decreased leaching losses (Sanchez, 1973b; De Datta and Kerim, 1974). In some reports, higher percolation rates have reduced water and nutrient economy and grain yields (Kira *et al.*, 1958; IRRI, 1966, 1967; Takai *et al.*, 1969; Kobo *et al.*, 1970; Matsura *et al.*, 1972; De Datta and Kerim, 1974; Shoji *et al.*, 1974; Ghildyal, 1982; Sharma and De Datta, 1985). Some data on the effect of puddling on soil permeability and leaching losses of nutrients at the IRRI farm are given in Table 6. However, under certain conditions, where phytotoxins accumulate in the root zone, some percolation may be essential for increasing rice grain yields. In temperate areas, because of low soil temperature, microbial decomposition of organic residues and organic acids is low (Yamane and Sato, 1967; Yoneyama and Yoshida, 1977; Savant and De Datta, 1982). Consequently, although organic acids are relatively slowly produced, they stay longer in soil (Yamane and Sato, 1967; Cho and Ponnamperuma, 1971). Rice growth is very sensitive to organic acids (Tadano and Yoshida, 1978). Unless they leach properly, rice growth and grain yields are reduced (Desai *et al.*, 1957; Hollis, 1967; Sugimoto, 1969; Cho and Ponnamperuma, 1971; Chandrasekaran and Yoshida, 1973; Govindasamy and Chandrasekaran, 1979; C.L. Patel *et al.*, 1979; C.L. Patel and Ghildyal, 1980). Higher percolation rates help in increasing rice grain yields primarily by keeping the phytotoxins in the root zone at low levels. In addition, higher percolation rates may benefit rice growth by regulating soil temperature and oxygen content of soil solution, thereby favoring root growth and nutrient uptake (Alva and Petersen, 1979; C.L. Patel *et al.*, 1984; Hasegawa *et al.*, 1985). Nevertheless, many reports suggest almost nonsignificant contribution of percolating water in improving oxygen environment of rice soils (Houng, 1981). Takai *et al.* (1974) reported that in soils containing large amounts of easily decomposable organic substances, higher percolation rates might decrease the soil redox potential instead of increasing it. The decrease in redox potential (Eh) results from the supply of molecular oxygen and removal of water-soluble substances toxic to microorganisms responsible for organic matter decomposition. Enhanced microbial activities increase the production of $[NH_4]^+$ and Fe^{2+} and decrease Eh. In soils containing small amounts of decomposable organic matter, most of the O_2 supplied through percolating water is consumed in the upper 0.5-cm part of the soil (Figure 8). Data in Figure 9, from a greenhouse experiment, also show that with as high as 0.40 $\mu m\ s^{-1}$ percolation, only about 2 cm of the topsoil layer was oxidized. Based on the results of several experiments, Takai *et al.* (1974) gave a general picture of the effects of percolation on the soil redox potential, as shown in Figure 10. Research in Japan, China, and Korea suggests a percolation rate of 0.05 to 0.25 $\mu m\ s^{-1}$ is essential for getting rice grain yields above 6 t ha^{-1} (De Datta, 1981; Ming-hua, 1981; Hasegawa *et al.*, 1985; Lal, 1985a). However, this may not be true in

Table 6. Effect of puddling on leaching losses of nutrients

Treatment	PR[a] (μm s^{-1})	Leaching losses (kg ha^{-1})						
		$[NO_3]^-$-N	$[NH_4]^+$-N	P	K	Fe	Mn	Zn
Clay								
Puddled	0.02	1.07	0.43	0.27	23	0.85	1.86	0.18
Nonpuddled	0.03	1.46	0.37	0.29	25	0.60	1.02	0.13
LSD[b]	—	0.31[c]	ns	ns	ns	ns	ns	ns
Clay loam								
Puddled	0.03	1.57	0.33	0.21	14	1.07	1.84	0.12
Nonpuddled	0.10	8.67	1.62	0.86	71	1.28	1.13	0.40
LSD[b]	—	2.61[d]	0.34[d]	0.19[d]	30[d]	ns	ns	0.07[d]

[a]PR = percolation rate.
[b]LSD = least significant difference.
[c] = significant at 5% level.
[d] = significant at 1% level.

Reproduced from Sharma and De Datta, *Soil Science Society of America Journal,* Vol. 49, pp. 1451–1457, 1985, by permission of the Soil Sci. Soc. Am.

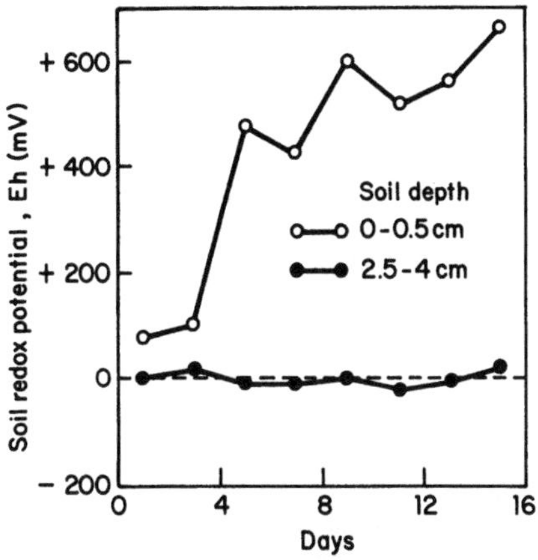

Figure 8. Effect of dissolved oxygen in percolating water on soil redox potential (Takai *et al.*, 1974). Reproduced with permission from Yasuo Takai, Nodai Research Institute, Tokyo University of Agriculture, Japan.

tropical areas, where fast microbial decomposition rates keep the organic matter and phytotoxins in soil at a low level. In a greenhouse experiment, the rice grain yields benefitted from percolation rates only in soils having organic matter above 50 g kg^{-1}. More research under controlled conditions is needed to establish the significance of higher percolation rates on rice grain yields in tropical areas.

6. Thermal Properties

Puddling affects the thermal properties of soil by changing soil properties, such as bulk density, moisture content, and percolation rates.

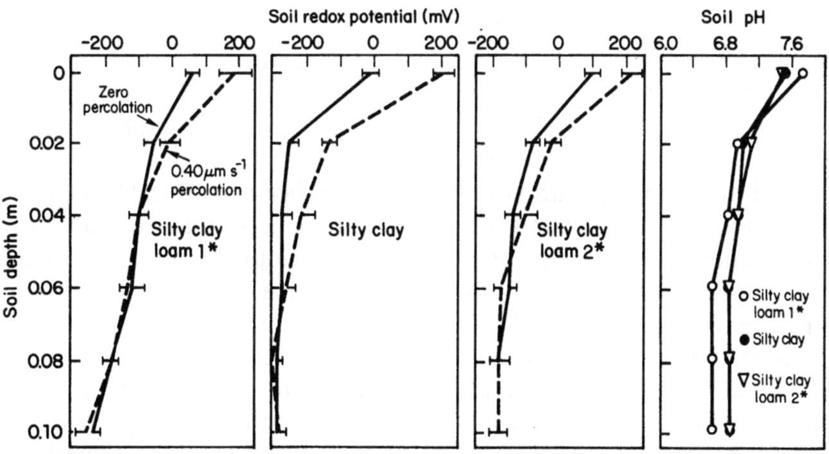

Figure 9. Effect of percolation on the redox potential of different soils (P.K. Sharma and S.K. De Datta, unpublished). *With different chemical properties.

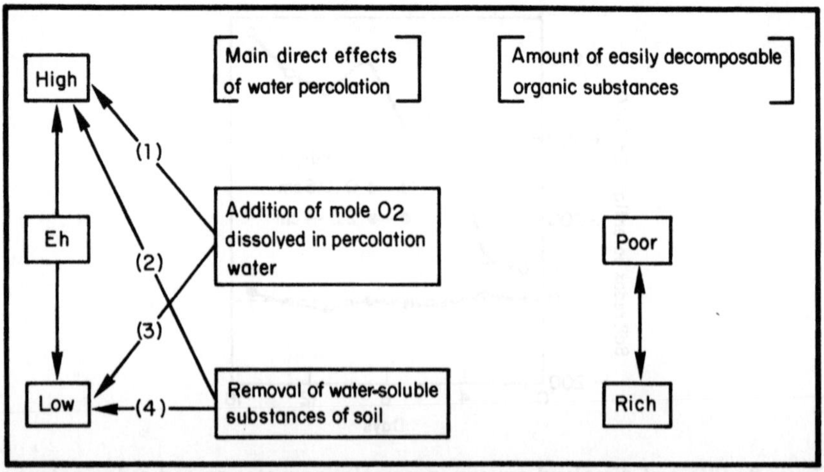

Figure 10. Effect of percolation of water on microbial metabolism in submerged soils. (1) Oxidation, (2) depression of bacterial activities by removal of substrates, (3) stimulation of aerobes' growth, (4) increase of bacterial activities by removal of toxic substances (Takai *et al.*, 1974). Reproduced with permission from Yasuo Takai, Nodai Research Institute, Tokyo University of Agriculture, Japan.

Thermal conductivity (k) and volumetric heat capacity (c) increase with bulk density and moisture content, because the k and c of soil particles and water are much higher than those of air (Ghildyal and Tripathi, 1971; Kasubuchi, 1975; Tripathi and Ghildyal, 1974; Hillel, 1980). Thermal diffusivity (k/c), which denotes the temperature change in any part of soil, also increases with increasing moisture content to about −0.1 MPa water potential and then it falls, because above −0.1 MPa water potential c increases much faster than k (Al Nakshabandi and Kohnke, 1965). According to Acharya and Gupta (1975), thermal properties are influenced more by moisture than by bulk density. Higher percolation rates may increase or decrease soil temperature depending on irrigation water temperature and solar radiation received. However, information regarding temperature profiles in lowland rice soils is completely lacking.

Some field and greenhouse studies have been conducted at IRRI, Philippines, to understand soil temperature responses to puddling, depth of flood water, and percolation rates in a tropical soil, where higher soil temperature may be a critical factor for rice production. Soil temperatures were recorded at hourly intervals and at different soil depths using copper (+) and constantan (−) thermocouples, and an Omega digital thermometer (Model 2175A). Maximum soil temperature and daily temperature amplitudes in 0- to 15-cm depth of a clay loam soil decreased with puddling (Figure 11). The effects of flood water depth on soil temperature are shown in Figure 12. Floodwater temperatures were highest with the lowest water depth. At 5-cm soil depth, maximum soil

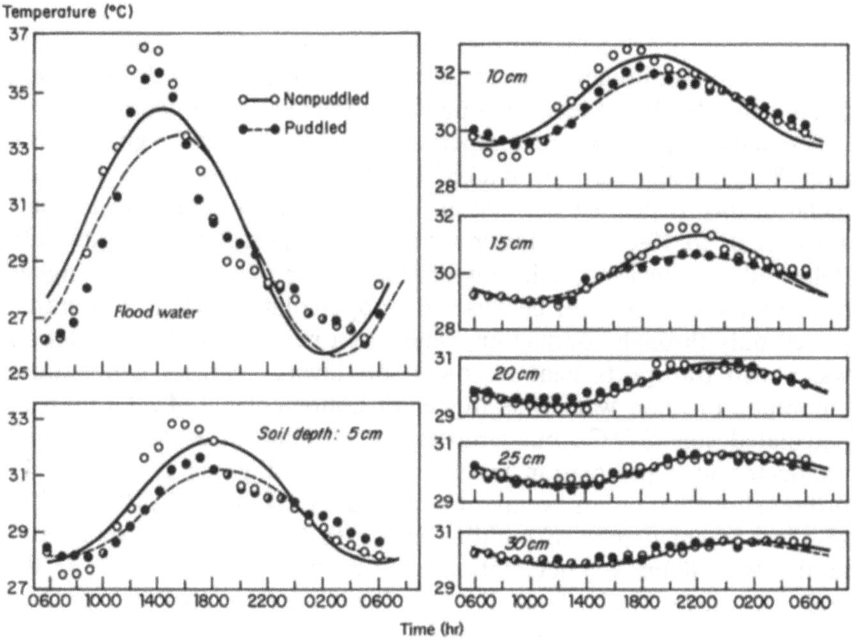

Figure 11. Effect of puddling on the diurnal temperature fluctuations in a clay loam soil (P.K. Sharma and S.K. De Datta, unpublished).

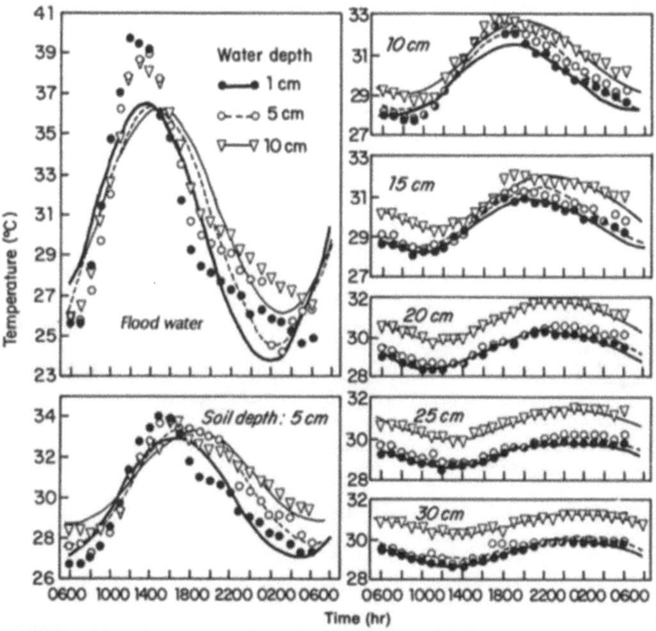

Figure 12. Effect of flood water depth on the diurnal temperature fluctuations in a clay loam soil (P.K. Sharma and S.K. De Datta, unpublished).

temperature under 1, 5, and 10 cm of flood water was the same, but below this depth deeper flood water maintained higher soil temperature. Daily temperature amplitudes in 0- to 10-cm depth decreased with increasing flood water depth but remained unaffected below 15-cm depth.

Percolation effects on soil temperature were studied in the greenhouse, with 90 kg of silty clay soil packed in 42-cm deep and 57-cm diameter metal drums. A percolation rate of 0.40 μm s^{-1} kept maximum temperature of 0- to 10-cm soil lower than without percolation (Figure 13). Below 10 cm, however, soil temperature of percolated soil was higher than nonpercolated soil, because in nonpercolated soil heat is transported only through conduction, whereas in percolated soil percolation water also transports heat to lower soil layers. Diurnal temperature amplitude of flood water was higher in nonpercolated (4.2°C) than in percolated soil (3.7°C), but temperature amplitudes in soil remained unaffected with percolation treatment. Data in Figures 11, 12, and 13 further show that on an average, temperature peaks at 0-, 5-, 10-, 15-, 20-, 25-, and 30-cm depths of puddled soil with 5 cm of flood water and moderate (<40 μm s^{-1}) percolation rate occurred at 14, 17, 19, 22, 23, 01, and 01 hr, respectively. Thus, the time lag between temperature peak at the soil surface (i.e., flood water) and at 30-cm soil depth was about 11 hr, as compared to less than 7 hr in an upland soil, as reported by B. Singh and Sandhu (1979) from north India.

These data lead to two important conclusions. First, puddling of

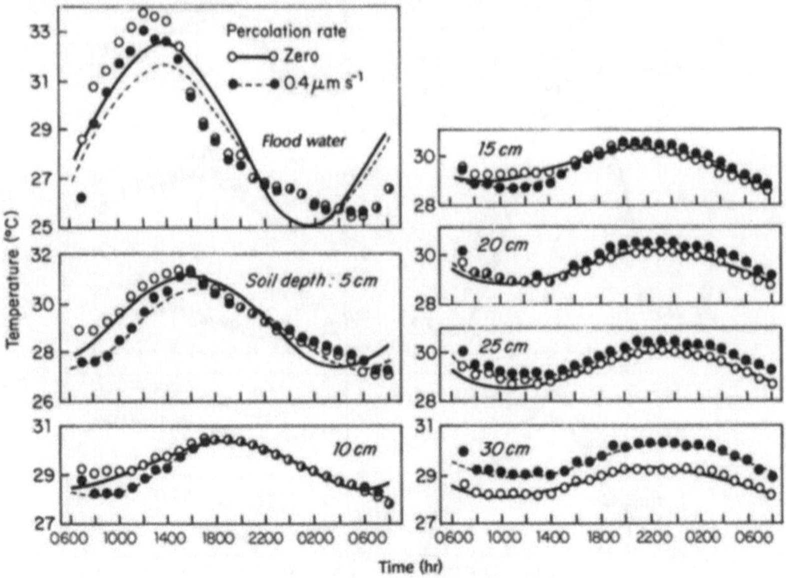

Figure 13. Effect of percolation on the diurnal temperature fluctuations in a silty clay soil (P.K. Sharma and S.K. De Datta, unpublished).

tropical soils decreases soil temperature and temperature amplitudes at least in the top 15-cm layer, where more than 80% of the rice roots are concentrated. Second, proper water management in rice fields is crucial for regulating proper soil thermal regimes. Shallow flood water is more susceptible to weather elements than deeper flood water. Consequently, relatively deep flood water in rice fields would buffer soils against extreme temperature fluctuations and help improve rice production. From Hungary Vamadevan (1971) reported greater plant height, panicle length, number of filled spikelets per panicle, 1000 grain weight, and grain yield with 20 cm than with 5 cm of flood water. Another report from a temperate region of India shows that ponded water in a rice field maintained soil temperature at 7.5-cm depth about 5°C higher than the continuously flowing water throughout the crop season and significantly increased the grain yield of rice (Kanwar *et al.*, 1978). Further, relatively high percolation will keep the maximum temperature of the root zone in tropical areas lower and the minimum temperature in temperate areas higher, depending on irrigation water temperature, thereby regulating soil temperature more favorably for rice growth. However, more research is needed to characterize puddled soils for their thermal properties under the different climatic conditions that affect rice growth from germination through root and shoot growth to the grain yield (Bhattacharya and De Datta, 1971; Cho and Ponnamperuma, 1971; Kar *et al.*, 1976; Chaudhary and Sandhu, 1982; Ghildyal, 1982).

B. Long-Term Effects

The long-term physical effect of puddling is the formation of hardpans in the subsoil, below the puddled layer. Depending on soil type, climate, hydrology, and frequency of puddling, a hardpan may form within as little as 3 years, as reported for fine sandy loams in Nigeria (Curfs, 1976) or as many as 200 years as in polder lands of the Shiroishi area, Kyushu, Japan (Kanno *et al.*, 1964; Motomura *et al.*, 1970). Figure 14 shows soil strength profiles from two texturally different lowland areas at an IRRI farm, which have been under wet tillage for 14 years.

Subsurface hard layers develop in two ways: (1) physical compaction and (2) chemical precipitation of Fe, Mn, and Si. Compact layers, about 5- to 10-cm thick, characterized by higher (dry) bulk density, lower total porosity, and lower water permeability than overlying and underlying soil horizons, occurring between 10- and 40-cm depth in lowland rice soils, have been described as "plow pans" by Koenigs (1950) and "traffic pans" by Moorman and van Breemen (1978). Tables 7 and 8 show some data on the physical properties of a hardpan formed in a lowland clay loam soil (Yun-sheng, 1983). These layers often form with the use of heavy machinery and from human and animal pressure during tillage, transplanting, weeding, etc. (Grant, 1965). Rozanov and Rozanova (1962)

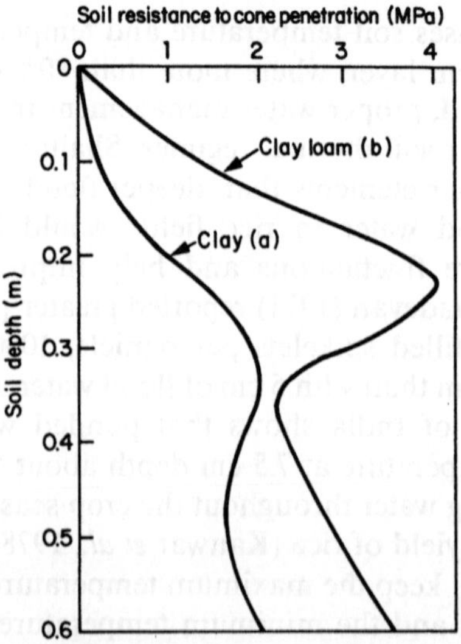

Figure 14. Penetrometer strength as a function of depth for two lowland soils at IRRI farm (a, data from P.K. Sharma and S.K. De Datta, unpublished; b, data from Dexter and Woodhead, 1985).

suggested that soil compaction might also result from periodic excessive wetting and intensive heating, alternating with drying of soil mass. "Traffic pans" are common in fine loamy soils (Mitsuchi, 1968). Very sandy and clayey soils often perform similarly.

Chemically cemented pans are formed by the precipitation of Fe, Mn, and Si, illuviated from reduced topsoil, to the oxidized subsoil, usually within 15- to 20-cm soil depth. Solubility of Fe and Mn is strongly influenced by soil redox potential. Reduction of Fe^{3+} to Fe^{2+} starts below 150 mV Eh and of Mn^{4+} to Mn^{2+} below 250 mV Eh (Patrick, 1981). As soon as appropriate redox potential is reached, Fe, Mn, and Si start migrating through the puddled soil downwards to the oxidized zone, where they get precipitated. Iron precipitates faster than Mn, because solubility of iron oxides is several times lower than that of manganese

Table 7. Bulk density (D_b), total porosity (n), aeration porosity at field capacity (n_a), and saturated hydraulic conductivity (K_s) of a typical lowland soil (Yun-sheng, 1983)

Soil property	*Plow layer*	*Plow pan*	*Below plow pan*	*Substratum*
D_b (Mg m^{-3})	1.20	1.35	—	1.56
n (%, v/v)	53	44	—	42
n_a (%, v/v)	15	2	3.7	Negligible
K_s (μm s^{-1})	120	0.19	0.36	0.12

Table 8. Pore-size distribution in a typical lowland soil (Yun-sheng, 1983)

Diameter of pores (mm)	Plow layer (29)[a]		Plow pan (28)[a]	
	(%, v/v)	% of total	(%, v/v)	% of total
>0.2	11.5 ± 3.1	22	5.5 ± 2.3	11
0.2 to 0.1	0.9 ± 0.5	2	0.4 ± 0.1	1
0.1 to 0.05	1.1 ± 0.3	2	0.5 ± 0.1	1
0.05 to 0.01	2.6 ± 0.9	5	1.7 ± 0.7	3
0.01 to 0.005	1.6 ± 0.4	3	1.3 ± 0.5	3
00.005	34.5 ± 3.1	66	40.9 ± 3.2	81

[a]Number of samples in parentheses.

oxides. Consequently, two distinct layers of precipitated Fe and Mn are observed in the subsoil of lowland rice (Koenigs, 1950; Kawaguchi and Matsuo, 1957; Kanno *et al.*, 1964). Soils with slowly permeable subsurface horizons, oxidized subsoil condition, low pH, high contents of easily reducible Fe and Mn, and easily decomposable organic matter offer better opportunities for chemical precipitation. In highly porous soils, instead of a continuous pan, a layer of iron concretions is formed (Springer, 1950; Dudal and Soepraptohardjo, 1963). In continuously submerged soils, because of excessively reduced conditions, the possibility of Fe and Mn pan formation is low. In highly weathered as well as acid sulfate soils, because of the low content of Mn oxides, only Fe pans may be formed without Mn accumulation (van Breemen, 1976; Kawaguchi and Kyuma, 1977). Under favorable conditions, an ironpan may form within 40 years and an Mn pan within 8 years (Kawaguchi and Matsuo, 1957).

Some workers associate ironpan formation with clay accumulations (Oades, 1963). Nagelschmidt (1939) suggested that kaolinite (>12%, w/w) in the soil fabric fixes iron oxides descending from the surface and provides a surface for pan formation. In most studies, however, clay migration has not been observed as a prerequisite for ironpan formation (Grant, 1965). In some lowland soils, weathering of clay minerals associated with alternate flooding and drainage is responsible for lower clay contents near the surface. This process, described as "ferrolysis" by Brinkman (1979), gradually lowers soil productivity.

Subsurface compact layers aid lowland rice because they cut down the water requirement and reduce leaching of nutrients. Mallick *et al.* (1976) reported that a dense layer with 1.65 Mg m^{-3} bulk density at 20-cm depth reduced water requirement of rice crop by 20 to 40%. Many reports relate the beneficial effects of artificial subsurface barriers to lowland rice production (R.P. Gupta and Nagarajarao, 1982). Plow pans help support traffic in lowland areas. Deep tillage may break such pans and make the soil deep, soft, and muddy, inaccessible to men, animals, and machines, as happened in northeastern Thailand (Moormann and van Breemen,

1978). However, plow pans also have disadvantages in restricting the proper root development and lowering grain yields, especially of upland crops. Deep tillage is needed in such cases to break hardpans and improve crop production (Carlson, 1978; Unger, 1979; R.P. Gupta *et al.*, 1984). Hardpans may also hurt rain-fed lowland rice under drought stress, but information is lacking in this respect. It is important to establish the relationship between subsurface soil compaction to different degrees and at different depths, pore-size distributions and the rigidity of the pore system in the hard layer, and root growth and grain yield of rain-fed lowland rice.

Furthermore, the subsurface compact layer has appeared in the literature by different names, e.g., plow pan, traffic pan, hard layer, compacted layer, restricting layer, impermeable layer. These names sometimes indicate the "cause" and sometimes the "effect" of these subsurface layers. It is essential to characterize these layers in terms of their location, rigidity, and porosity and accordingly give them a suitable, common name.

V. Regeneration of Soil Structure

Although puddling is considered beneficial for rice for various reasons, it can hurt upland crops by deteriorating soil structure and reducing soil permeability. The regeneration of soil structure is therefore relevant where an upland crop follows lowland rice.

Many soils develop friable structure on drying (Jewitt *et al.*, 1979), but natural structural regeneration is slow. The speed of recovery of soil structure depends on soil texture, clay mineralogy, organic matter, and environmental conditions. Regeneration is faster in kaolinitic clay soils rich in organic matter and iron oxides than in soils dominated by montmorillonitic clays and poor organic matter (Briones, 1977). Regeneration can be speeded up by adopting suitable tillage techniques at appropriate moisture contents and proper organic matter management.

Tillage is difficult in clay soils because they are hard when dry and sticky when wet. Although very little is known about the workability of these soils and about soil-implement interactions, puddling, in general, has been found to decrease the soil moisture range over which an optimum soil tilth can be obtained. According to Utomo and Dexter (1981), for some soils, a gravimetric moisture content around 0.9 times the plastic limit provides maximum friability. At this moisture content, soil tillage maximizes the proportion of small aggregates (Ojeniyi and Dexter, 1979). However, one problem with lowland soils is that they dry slowly, often taking weeks to months to reach a workable moisture content. This increases the turnaround time between two crops (Syarifuddin, 1982; Morris, 1982). Upon drying, puddled soils become impenetratable to roots because of increased mechanical impedance (Taylor *et al.* 1964; Syarifuddin and Zandstra, 1978). With excess moisture, it may be worthwhile to try zero or minimum tillage for upland crops after lowland rice, with proper crop selection (Khan *et al.*, 1981; Zandstra, 1982). If a crop can be grown satisfactorily by reduced tillage, it will shorten the turnaround time and also alleviate problems in creating an adequate seedbed. However, without tillage, at high moisture contents upland crops after lowland rice on poorly drained soils may suffer from anaerobic conditions (Gill and Miller, 1956; Wien *et al.*, 1979). The use of oxygenating compounds, such as calcium peroxide, has been found useful in improving emergence and grain yields of crops (Callebaut *et al.*, 1981; Ogunremi *et al.*, 1981). These compounds may also help in removing the deleterious effects of temporary water stagnations due to occasional rains on upland crops. The problem of anaerobiosis can also be alleviated by tilling the surface soil. In a field experiment, under a shallow water table (0.1 to 0.3 m) condition, which kept soil water potential at 10-cm depth below −0.007 MPa, shallow tillage (7- to 10-cm deep) with a rotovator increased grain yield of maize (*Zea mays* L.) by about 230%, and of mung bean (*Vigna radiata* L.) by about 25% over yields without tillage. The increase in yield was attributed to the increased aeration porosity in the top 10-cm layer of tilled soil. Sorghum [*Sorghum bicolor* (L.) Moench.], on the other hand, did not respond to tillage in soil with a 50- to 100-cm water table (Hundal and De Datta, 1984b). These data suggest that for crops sensitive to poor soil aeration, such as maize, shallow tillage is better than zero tillage and can be performed at relatively high soil moisture contents.

Land management and soil water conservation during a dry season fallow period help in growing two crops of rice during the wet season by advancing the establishment of first crop by 3 to 4 weeks (Bolton and De Datta, 1979; Hundal and De Datta, 1982). This technique may help increase the cropping intensity in rain-fed areas adopting rice-based cropping systems.

Addition of rice straw and compost to rice crop has been found to improve aggregation in puddled soils (Chaudhary and Ghildyal, 1969; Sahoo *et al.*, 1970). Inclusion of legumes, such as chickpeas (*Cicer arietinum* L.), in rotation with lowland rice improved the structure of lowland soil (Padma Raju and Deb, 1969). According to Prasad *et al.* (1980), broadcasting sprouted seeds on puddled soil resulted in less soil cracking and better aggregation than transplanting on a puddled soil or direct seeding on a moist soil. This probably resulted from the positioning of plants in the soil, as has been reported for other crops, such as wheat, in vertisols (Johnson, 1962; Fox, 1964).

VI. Is Puddling Essential?

Because puddling requires intensive labor and capital and creates soil physical conditions detrimental to upland crops in rice-based cropping systems, it is logical to analyze whether puddling is really essential in rice culture. It is also logical to identify conditions where less intensive tillage can substitute for puddling. Research shows that in many cases, zero and minimum tillage have produced rice grain yields similar to those with conventional tillage of puddling (Mittra and Pieris, 1968; Mabbayad and Buencosa, 1967; Mabbayad *et al.*, 1968; Seth *et al.*, 1971; N.T. Singh *et al.*, 1977; Croon, 1978; De Datta *et al.*, 1979; Maurya and Lal, 1979; Rodriguez and Lal, 1979). Lal (1985a) has described in detail the responses of rice to wet or dry tillage under different soil conditions. Puddling has been found to benefit rice only on those soils which undergo drastic modifications in their physical properties due to wet tillage. For example, in clayey soils with well-developed and water-stable aggregates and having high permeability, puddling is rather essential for lowland rice. However, in clayey soils dominated by active (2:1) clays, which disperse easily on wetting, and soils having low permeability, rice can be grown successfully even without puddling.

If puddling is necessary, its frequency can be reduced. Data shown in Table 9 and some earlier reports (Mabbayad and Buencosa, 1967; Croon, 1978) support this conclusion. However, weeds are the major problem associated with zero or minimum tillage. This problem can be tackled by combining herbicides with proper soil and crop management (De Datta

Table 9. Response of lowland rice (IR36) to methods of land preparation in two different soils

	Grain yield (t ha^{-1})	
Treatment	*Soil I*	*Soil II*
Zero tillage	4.3	3.9
Minimum tillage[a]	5.5	4.4
Conventional tillage[b]	5.5	5.5
LSD (5%)	0.5	0.9
Texture	Clay	Clay loam
Clay[c] (%)	53	28
Dominant clay mineral	Montmorillonite-like	Halloysite-like
Degree of shrinkage (%, v/v)	47	30
Percolation and seepage (μm s^{-1})	0.4	10.4
Water-table depth (m)	0.2	>1.5

[a]One dry rotovation + submergence.

[b]Two wet plowings + three harrowings.

[c]About 90% of the clay fraction in both soils is amorphous aluminosilicates.

P.K. Sharma and S.K. De Datta, unpublished data.

et al., 1979). On coarse- to medium-textured soils with low activity (1:1) clays, having a single grain to poorly developed structure and high permeability, and which respond little to wet-tillage operations, puddling does not benefit rice (FAO, 1976; Mittra and Pieris, 1968; Maurya and Lal, 1979; Lal, 1985a; Rodriguez and Lal, 1985). For such areas, Ghildyal (1969) suggested soil compaction as an alternative to puddling. Moderate soil compactions improved rice yields by reducing percolation losses and favoring root–soil contact (Varade and Ghildyal, 1967; Kumar *et al.*, 1971; Kar *et al.*, 1976; Ghildyal, 1978; M.S. Patel and Singh, 1979; N.T. Singh *et al.*, 1980; Ogunremi *et al.*, 1985). On other soils where permeability is low and phytotoxins might accumulate in the root zone, as in floodplain soils, dry tillage may be a better choice than wet tillage. Lal (1985b) developed a rating system to assess soil tillage requirements for lowland rice-based cropping systems, based on such parameters as cation exchange capacity (CEC), soil permeability, endurance to tillage, and days available for seedbed preparation. The system needs to be tested under different soil and climatic conditions.

VII. Research Gaps

Although considerable information is available on puddling effects on the growth and yield of rice and its consequences in rice-based cropping systems, some areas need more study:

1. Identification of soil properties and cropping systems where puddling is not the only choice in rice production.
2. Development of suitable indices to characterize the degree of puddling.
3. Effect of percolation rates on nutrient and water economy, soil redox potential, and removal of phytotoxins from root zone under different soil conditions. (This has a special relevance in areas where organic manures, green manures, and the like, are used as an alternative or supplemental source to chemical fertilizers.)
4. Soil thermal regimes of puddled soils in relation to rice production.
5. Effect of subsurface hard layers, varying in pore-size distribution and soil strength, on rain-fed lowland rice.
6. Relative performance of dry-seeded and transplanted rice in a rain-fed situation, when water is a limiting factor.
7. Long-term evaluation of minimum tillage compared to puddling in terms of modifications in soil physical environment and production of lowland rice and following rain-fed upland crops.
8. Alternative tillage techniques for upland crops following lowland rice with the objective of reducing turnaround time between two crops in rice-based cropping systems.

VIII. Summary

This contribution has reviewed the puddling processes and their consequences on soil physical properties in relation to crop production in lowland rice-based cropping systems. Physical edaphic factors, such as soil aggregation, bulk density, soil strength, pore-size distributions, gaseous exchange, water retention and transmission, soil thermal regimes, and the formation of subsurface hardpans have been discussed in relation to puddling.

Puddling destroys soil aggregates, triggering changes in other soil physical properties. Bulk density may increase or decrease depending on the structural status of the soil before puddling. Soil strength decreases with puddling. Puddling eliminates water-transmission pores and increases storage and residual pores. Puddling drastically reduces gaseous exchange. Hydraulic conductivity, percolation rate, and evaporation loss are minimized upon puddling. Puddling increases water retention below −0.01 MPa water potential. It lowers soil temperature of the root zone in tropical areas. The influences of these modifications on rice production have been discussed. The report further has dealt with hardpan formation in lowland rice soils below puddled layers and their significance in crop production in rice-based cropping systems. Finally, some unresolved questions, considering rice-based cropping system as a whole, have been indicated.

References

Acharya, C.L., and R.P. Gupta. 1975. Thermal diffusivity values based upon time dependent soil temperature distributions. *J. Indian Soc. Soil Sci.* 23:1–7.

Agarwal, M.C., R. Singh, M.L. Batra, and R.P. Agrawal. 1978. Evaluation of different implements for puddling of rice soils. *Il Riso* 27:301–305.

Allman, M.S., and H. Kohnke. 1947. The pF of soil moisture at the wet limit of the plowing range. *Soil Sci. Soc. Am. Proc.* 12:22–23.

Al Nakshabandi, G., and H. Kohnke. 1965. Thermal conductivity and diffusivity of soils as related to moisture tension and other physical properties. *Agr. Meteorol.* 2:271–279.

Alva, A.K., and L. Petersen. 1979. Soil physical properties in relation to rice yield and water consumption under flooded and unflooded conditions. *Plant and Soil* 52:353–363.

Aomine, S., and Y. Shiga. 1959. Soil fabrics of the plowed layer of flooded rice fields. *Soil Plant Food* 5:64–72.

Asthana, D.C., and V.K. Vamadevan. 1977. Note on water losses in rice culture. *Indian J. Agri. Sci.* 47:216–217.

Aylmore, L.A.G., and J.P. Quirk. 1962. The structural status of clay systems. *Clay and Clay Miner.* 9:104–130.

Baver, L.D. 1956. *Soil physics*. John Wiley & Sons, New York.

Beacher, B.F., and E. Strickling. 1955. Effect of puddling on water stability and bulk density of aggregates of certain Maryland soils. *Soil Sci.* 80:363–373.

Bhattacharyya, A.K., and S.K. De Datta. 1971. Effects of soil temperature regimes on growth characteristics, nutrition and grain yield of IR22 rice. *Agron. J.* 63:443–449.

Bhole, N.G. 1963. Measurement of quality of puddle. M. Tech. Thesis, Indian Inst. Tech., Kharagpur, India (unpublished).

Bodman, G.B., and P.R. Day. 1943. Freezing points of a group of California soils and their extracted clays. *Soil Sci.* 55:224–246.

Bodman, G.B., and J. Rubin. 1948. Soil puddling. *Soil Sci. Soc. Am. Proc.* 13:27–36.

Bolton, F.R., and S.K. De Datta. 1979. Dry soil mulching in tropical rice. *Soil Sci. Plant Nutr.* 25:173–181.

Brinkman, R. 1979. *Ferrolysis, a soil-forming process in hydromorphic conditions.* Centre for Agricultural Publishing and Documentation, Wageningen, Netherlands, pp. 106.

Briones, A.A. 1977. Aggregation in drained and reclaimed paddies. Dept. of Soils, Coll. Agr., Univ. Philipp., Los Baños, Laguna, Philippines (unpublished).

Briones, A.A., and M.E. Raymundo. 1962. The influence of varying depths of mud on the yield of lowland rice. *Phil. Agr.* 45:567–572.

Buehrer, T.F., and M.S. Rose. 1943. Studies in soil structure. V. Bound water in normal and puddled soils. *Ariz. Agr. Exp. Sta. Tech. Bull.* 100:155–218.

Callebaut, F., D. Gabriels, and M. De Boodt. 1981. Time responses of sugarbeet germination, oxygen diffusion, and redox potential to crust formation, polyacrylamide stabilization and peroxide treatment. *Geoderma* 25:275–283.

Campbell, R.B. 1952. Freezing point of water in puddled and unpuddled soils at different soil moisture tension values. *Soil Sci.* 73:221–229.

Carlson, C.W. 1978. Research in ARS related to soil structure. In: W.W. Emerson, R.D. Bond, and A.R. Dexter (eds.), *Modification of soil structure.* John Wiley and Sons, New York, pp. 279–284.

Cernuda, C.F., R.M. Smith, and J. Vicente-Chandler. 1954. Influence of initial soil moisture conditions on resistance of macroaggregates to slaking and to water-drop impact. *Soil Sci.* 77:19–27.

Chandrasekaran, S., and T. Yoshida. 1973. Effect of organic acid transformations in submerged soils on growth of the rice plant. *Soil Sci. Plant Nutr.* 19:39–45.

Chang, C.W. 1941. An experimental study on the development of the adobe structures in soils. *Soil Sci.* 52:213–228.

Chaudhary, T.N., and B.P. Ghildyal. 1969. Aggregate stability of puddled soil during rice growth. *J. Indian Soc. Soil Sci.* 17:261–265.

Chaudhary, T.N., and B.S. Sandhu. 1982. Soil temperature and plant growth. In: *Review of Soil Research in India Part I, 12th Int. Congr. Soil Sci.* Indian Society of Soil Science, New Delhi, India, pp. 48–59.

Cho, D.Y., and F.N. Ponnamperuma. 1971. Influence of soil temperature on the chemical kinetics of flooded soils and the growth of rice. *Soil Sci.* 112:184–194.

Croney, D., and J.D. Coleman. 1954. Soil structure in relation to soil suction (pF). *J. Soil Sci.* 5:75–84.

Croon, F.W. 1978. Zero-tillage for rice on vertisols. *World Crops Livestock* 30:12–16.

Curfs, H.P.F. 1976. *Systems development in agricultural mechanization with special reference to soil tillage and weed control*. Mededelingen Landbouwhogeschool, Wageningen, Netherlands 76–5.

De Datta, S.K. 1981. *Principles and practices of rice production*. John Wiley & Sons, New York.

De Datta, S.K., and R. Barker. 1978. Land preparation for rice soils. In: *Soils and rice*. Int. Rice Res. Inst., Los Baños, Philippines, pp. 623–648.

De Datta, S.K., and M.S.A.A.A. Kerim. 1974. Water and nitrogen economy of rainfed rice as affected by soil puddling. *Soil Sci. Soc. Am. Proc.* 38:515–518.

De Datta, S.K., H.K. Krupp, E.I. Alvarez, and S.C. Modgal. 1973. Water management practices in flooded tropical rice. In: *Water management in Philippine irrigation systems: research & operations*. Int. Rice Res. Inst., Los Baños, Philippines, pp. 1–18.

De Datta, S.K., F.R. Bolton, and W.L. Lin. 1979. Prospects for using minimum, and zero tillage in tropical lowland rice. *Weed Res.* 19:9–15.

De Jong, E., and E.A. Paul. 1979. Aeration, respiration and atmosphere. In: R.W. Fairbridge and C.W. Finkl, Jr.(eds.), *The encyclopedia of soil science, Part I*. Dowden, Hutchinson & Ross, Inc., Stroudsburg, Pennsylvania, pp. 10–21.

Desai, A.D., T. Seshagiri Rao, and L.R. Hirekerur. 1957. Drainage for rice fields. Sulphide toxicity. *J. Indian Soc. Soil Sci.* 5:155–161.

Dexter, A.R., and T. Woodhead. 1985. Soil mechanics in relation to tillage, implements, and root penetration in lowland soils. In: *Soil physics and rice*. Int. Rice Res. Inst., Los Baños, Philippines, pp. 261–281.

Dudal, R., and M. Soepraptohardjo. 1963. Soil classification in Indonesia. Gen. Agr. Res. Sta. Bull. 148, p. 16.

Dutt, A.K. 1948. Puddling and other treatments in relation to soil structure and crop growth. *J. Am. Soc. Agron.* 20:243–330.

Elrick, D.E., and C.B. Tanner. 1955. Influence of sample pretreatment on soil moisture retention. *Soil Sci. Soc. Am. Proc.* 19:279–282.

Fagi, A.M., and S.K. De Datta. 1983. Physical properties of rainfed wetland rice soils as affected by cropping system and crop residue management. *Field Crops Res.* 6:189–204.

FAO (Food and Agriculture Organization). 1976. *Mechanization of rice production*. An International Co-ordinated Research Project, FAO, Rome Italy.

Fox, W.E. 1964. Cracking characteristics and field capacity in a swelling soil. *Soil Sci.* 98:413.

Ghildyal, B.P. 1969. Soil compaction—a new system of rice cultivation. *Indian Farming* 19:19–22.

Ghildyal, B.P. 1978. Effects of compaction and puddling on soil physical properties and rice growth. In: *Soils and rice*. Int. Rice Res. Inst., Los Baños, Philippines, pp. 317–336.

Ghildyal, B.P. 1982. Nature, physical properties and management of submerged rice soils. In: *Vertisols and Rice Soils of the Tropics, Symposia Papers II, 12th Int. Congr. Soil Sci.* Indian Society of Soil Science, New Delhi, India, pp. 121–142.

Ghildyal, B.P., and T. Satyanarayana. 1965. Effect of compaction on physical properties of four different soils of India. *J. Indian Soc. Soil Sci.* 13:149–155.

Ghildyal, B.P., and R.P. Tripathi. 1971. Effect of varying bulk densities on the thermal characteristics of lateritic sandy clay loam soil. *J. Indian Soc. Soil Sci.* 19:5–10.

Gill, W.O., and R.D. Miller. 1956. A method for study of the influence of mechanical impedance and aeration on the growth of seedling roots. *Soil Sci. Soc. Am. Proc.* 20:154–157.

Govindasamy, R., and S. Chandrasekaran. 1979. Inhibitory effect of rice straw incorporation in ill-drained (Annamalai clay) soil on rice. *J. Indian Soc. Soil Sci.* 27:92–94.

Grant, C.J. 1965. Soil characteristics associated with the wet cultivation of rice. In: *The mineral nutrition of the rice plant.* John Hopkins University Press, Baltimore, pp. 15–28.

Greenland, D.J. 1981. Recent progress in studies of soil structure, and its relation to properties and management of paddy soils. In: Institute of Soil Science, Academia Sinica (ed.), *Proc. Symp. on Paddy Soil.* Science Press, Beijing, China, pp. 42–58.

Gupta, R.K., and I.K. Jaggi. 1979. Soil physical conditions and paddy yield as influenced by depth of puddling. *J. Agron. Crop Sci.* 148:329–336.

Gupta, R.P., and Y. Nagarajarao. 1982. Soil structure and its management. In: *Review of Soil Research in India, Part I, 12th Int. Congr. Soil Sci.* Indian Society of Soil Science, New Delhi, India, pp. 60–76.

Gupta, R.P., S. Kumar, and T. Singh (eds.). 1984. *Soil management to increase crop production.* Indian Council of Agricultural Research, New Delhi, India.

Harris, R.F., G. Chesters, and O.N. Allen. 1966. Dynamics of soil aggregation. *Adv. Agron.* 18:107–169.

Hasegawa, S., M. Thangaraj, and J.C. O'Toole. 1985. Root behavior: field and laboratory studies for rice and nonrice crops. In: *Soil physics and rice.* Int. Rice Res. Inst., Los Baños, Philippines, pp. 383–395.

Hillel, D. 1980. *Fundamentals of soil physics.* Academic Press, New York.

Hollis, J.P. 1967. Toxicant diseases of rice. Bull. Louisiana Agr. Exp. Stn. No. 614, 21 pp.

Houng, K.H. 1981. A theoretical evaluation of the influence of percolation rate on the thickness of the oxidizing zone of paddy soils. *Proc. Nat. Sci. Counc. Part A: Appl. Sci.* 5:274–278.

Huang, M.H. 1982. The identification of soil physical properties related to the growth and yield of lowland rice. *J. Agr. Res. China* 31:347–352.

Hundal, S.S., and S.K. De Datta. 1982. Effect of dry-season soil management of water conservation for the succeeding rice crop in a tropical soil. *Soil Sci. Soc Am. J.* 46:1081–1086.

Hundal, S.S., and S.K. De Datta. 1984a. In situ water transmission characteristics of a tropical soil under rice-based cropping systems. *Agr. Water Management* 8:387–396.

Hundal, S.S., and S.K. De Datta. 1984b. Water table and tillage effects on root distribution, soil water extraction, and yield of sorghum grown after wetland rice in a tropical soil. *Field Crops Res.* 9:291–303.

IRRI (International Rice Research Institute). 1966. *Annual Report for 1965.* Int. Rice Res. Inst., Los Baños, Philippines.

IRRI (International Rice Research Institute). 1967. *Annual Report for 1966.* Int. Rice Res. Inst., Los Baños, Philippines.

IRRI (International Rice Research Institute). 1973. *Annual Report for 1972.* Int. Rice Res. Inst., Los Baños, Philippines.

Jamison, V.C. 1953. Changes in air-water relationships due to structural improvement of soils. *Soil Sci.* 76:143–151.

Jewitt, T.N., R.D. Law, and K.J. Virgo. 1979. Vertisol soils of the tropics and subtropics: their management and use. *Outlook on Agriculture* 10:33–40.

Johnson, W.C. 1962. Controlled soil cracking as a possible means of moisture conservation on wheat lands of the Southwestern Great Plains. *Agron. J.* 54:323–325.

Kanno, I., Y. Honyo, S. Arimura, and S. Tokudome. 1964. Genesis and characteristics of rice soils developed on polder lands of Shiroishi area, Kyushu. *Soil Sci. Plant Nutr.* 10:1–20.

Kanwar, B.S., S.S. Masand, and B.R. Tripathi. 1978. Effect of continuously flowing and ponded *Kuhl* water application on performance of rice in Himachal Pradesh. *J. Indian Soc. Soil Sci.* 26:257–260.

Kar, S., and B.P. Ghildyal. 1975. Rice root growth in relation to size, quantity and rigidity of pores. *Plant Soil* 43:627–637.

Kar, S., and S.B. Varade. 1972. Influence of mechanical impedance on rice seedling root growth. *Agron. J.* 64:80–81.

Kar, S., S.B. Varade, T.K. Subramanyam, and B.P. Ghildyal. 1976. Soil physical conditions affecting rice root growth: bulk density and submerged soil temperature regime effects. *Agron. J.* 68:23–26.

Kar, S., S.B. Varade, and B.P. Ghildyal. 1979a. Pore size distribution and root growth relations of rice in artificially synthesized soils. *Soil Sci.* 128:364–368.

Kar, S., S.B. Varade, and B.P. Ghildyal. 1979b. Soil physical environment affecting root growth of upland rice. *J. Agri. Sci. Cambridge* 93:719–726.

Kasubuchi, T. 1975. The effect of soil moisture on thermal properties in some typical Japanese upland soils. *Soil Sci. Plant Nutr.* 21:107–112.

Kawaguchi, K., and D. Kita. 1957. Mechanical and chemical constituents of water-stable aggregates of paddy soil with relationship to the aggregate size. *Soil Plant Food* 3:22–28.

Kawaguchi, K., and K. Kyuma. 1977. *Paddy soils in Tropical Asia: their material, nature and fertility*. The University Press of Hawaii, Honolulu.

Kawaguchi, K., and Y. Matsuo. 1957. Re-investigation on distribution of active and inactive oxides along soil profiles in time series of dry rice fields in polder lands of Kojima basin, Okayama Prefecture, Japan. *Soil Plant Food* 3:29–35.

Kawaguchi, K., D. Kita, and K. Kyuma. 1956. A soil core sampler for paddy soils and some physical properties of the soils under water logged condition. *Soil Plant Food* 2:92–95.

Khan, S.A., A.S. Roy, and B.N. Chatterjee. 1981. Note on growing winter maize under minimum tillage after transplanted rice. *Indian J. Agr. Sci.* 51:54–55.

Kira, Y., K. Shuna, and H. Takenaka. 1958. The influence of percolation on paddy soil and rice plants. *J. Agr. Eng. Soc. Jpn.* 25:339–344.

Kisu, M. 1978. Tillage properties of wet soils. In: *Soils and rice*. Int. Rice Res. Inst., Los Baños, Philippines, pp. 307–316.

Kobo, K., Y. Takai, H. Wada, and H. Kagawa. 1970. Effects of percolation on the chemical and biological condition of water-logged paddy soils. I. Oxidation-reduction potential. *J. Sci. Soil Manure Jpn.* 41:438–444; also in *Soil Sci. Plant Nutr.* 17:174 (Engl. Abstr.).

Koenigs, F.F.R. 1950. A "sawah" profile near Bogor (Java). *Trans. IV Int. Congr. Soil Sci.* 1:297–300.

Koenigs, F.F.R. 1961. *The mechanical stability of clay soils as influenced by the moisture conditions and some other factors*. Wageningen, Verslagen van Landbounkundige Onderzockingen.

Koenigs, F.F.R. 1963. The puddling of clay soils. *Neth. J. Agr. Sci.* 11:145–156.

Kokubun, K., H. Masujima, and K. Nemoto. 1969. Effects of machinery use on permeability and structure of soil in paddy fields. *J. Sci. Soil Manure Jpn.* 40:276–280; also in *Soil Sci. Plant Nutr.* 16:175 (Engl. Abstr.).

Kumar, V., K.T. Mahajan, S.B. Varade, and B.P. Ghildyal. 1971. Growth response of rice (*Oryza sativa* L.) to submergence, soil aeration and soil strength. *Indian J. Agr. Sci.* 41:527–534.

Lachenbruch, A.H. 1961. Depth and spacing of tension cracks. *J. Geophys. Res.* 66:4273–4292.

Lal, R. 1985a. Tillage in lowland rice-based cropping systems. In: *Soil physics and rice.* Int. Rice Res. Inst., Los Baños, Philippines, pp. 283–307.

Lal, R. 1985b. A soil suitability guide for different tillage systems in the tropics. *Soil Till. Res.* 5:179–196.

Letey, J., and L.H. Stolzy. 1964. Measurement of oxygen diffusion rates with the platinum microelectrode. III. Correlation of plant response to soil oxygen diffusion rates. *Hilgardia* 35:545–576.

Luxmoore, R.J., L.H. Stolzy, and J. Letey. 1970. Oxygen diffusion in the soil-plant system. 1. A model. *Agron. J.* 62:317–322.

Mabbayad, B.B., and I.A. Buencosa. 1967. Tests on minimal tillage of transplanted rice. *Phil. Agr.* 51:541–551.

Mabbayad, B.B., B.N. Emerson, and E.L. Aragon. 1968. Further tests on minimal tillage and rates of nitrogen application on transplanted rice. *Phil. Agr.* 52:200–210.

Mallick, S., T.V. Rao, and Y. Nagarajarao. 1976. Effect of subsurface compaction and bentonite application on the irrigation requirement and growth of rice. *Indian J. Agron.* 21:317–318.

Matsura, K., K. Matsubara, and Y. Sakanoue. 1972. Influence of percolation on the leaching of some mineral nutrients from water-logged paddy soils. *J. Sci. Soil Manure Jpn.* 43:238–244; also in *Soil Sci. Plant Nutr.* 20:197 (Engl. Abstr.).

Maurya, P.R., and R. Lal. 1979. Influence of tillage and seeding methods on flooded rice. In: R. Lal (ed.), *Soil tillage and crop production.* Proc. Series No. 2, IITA, Ibadan, Nigeria, pp. 337–347.

McGeorge, W.T. 1937. Studies on soil structure: some physical characteristics of puddled soils. *Ariz. Agr. Exp. Sta. Tech. Bull.* 67:127–177.

Ming-hua, F. 1981. Characteristics of high-yield paddy soils in suburbs of Shanghai. In: Institute of Soil Science, Academia Sinica (ed.), *Proc. Symp. on Paddy Soil.* Science Press, Beijing, China, pp. 769–774.

Mitsuchi, M. 1968. Profile differentiation of surface water type paddy soils under different drainage conditions. *J. Sci. Soil Manure Jpn.* 39:233–238; also in *Soil Sci. Plant Nutr.* 14:207 (Engl. Abstr.).

Mittra, M.K., and J.W.L. Pieris. 1968. Paraquat as an aid to paddy cultivation. *Proc. 9th Br. Weed Control Conf.*, London, United Kingdom, 9:668–674.

Moormann, F.R., and N. van Breemen. 1978. Soil forming processes in aquatic rice lands. In: *Rice: soil, water, land.* Int. Rice Res. Inst., Los Baños, Philippines, pp. 83–106.

Morris, R.A. 1982. Tillage and seeding methods for dry-seeded rice. In: *Report of a workshop on cropping systems research in Asia.* Int. Rice Res. Inst., Los Baños, Philippines, pp. 117–131.

Motomura, S., E.M. Lapid, and H. Yokoi. 1970. Soil structure development in Ariake polder soils in relation to iron forms. *Soil Sci. Plant Nutr.* 16:47–54.

Nagelschmidt, G. 1939. Identification of minerals in soil colloids. *J. Agr. Sci.* 29:477–501.

Naphade, J.D., and B.P. Ghildyal. 1971. Effect of puddling on physical properties of rice soil. *Indian J. Agr. Sci.* 41:1065–1067.

Oades, J.M. 1963. The nature and distribution of iron compounds in the soils. *Soils Fert.* 26:69–80.

Obermueller, A.J., and D.S. Mikkelsen. 1974. Effects of water management and soil aggregation on the growth and nutrient uptake of rice. *Agron. J.* 66:627–632.

Ogunremi, L.T., R. Lal, and O. Babalola. 1981. Effects of water table depth and calcium peroxide application on cowpea (*Vigna unguiculata*) and soybean (*Glycine max*). Plant Soil 63:275–281.

Ogunremi, L.T., R. Lal, and O. Babalola. 1985. Effects of tillage methods and water regimes on soil properties and yield of lowland rice from a sandy loam soil in southwest Nigeria. *Soil Till. Res.* (in press).

Ojeniyi, S.O., and A.R. Dexter. 1979. Soil factors affecting the macro-structures produced by tillage. *Trans. Am. Soc. Agr. Eng.* 22:339–343.

Padma Raju, A., and A.R. Deb. 1969. Influence of crop rotations on the structure of paddy soils. *Indian J. Agr. Sci.* 39:81–87.

Pande, H.K. 1975. Water management practices and rice cultivation in India. In: *Proc. Symp. Water Management in Rice Field.* Tropical Agriculture Research Center, Ibaraki, Japan, pp. 231–249

Patel, C.L., and B.P. Ghildyal. 1980. Effect of different drainage conditions on growth of rice. *Indian J. Agr.* 25:342–347.

Patel, C.L., B.P. Ghildyal, and V.S. Tomar. 1979. Water use, nutrient balance and growth of rice in different drainage conditions. *Il Riso* 28:291–296.

Patel, C.L., B.P. Ghildyal, and V.S. Tomar. 1984. Nutrient inflow rates in rice roots under varying drainage conditions. *Plant Soil* 77:243–252.

Patel, M.S., and N.T. Singh. 1979. The effect of soil compaction on growth and water use efficiency of rice. *Indian J. Agron.* 24:429–431.

Patrick, W.H. Jr. 1981. The role of inorganic redox systems in controlling reduction in paddy soils. In: Institute of Soil Science, Academia Sinica (ed.), *Proc. Symp. on Paddy Soil.* Science Press, Beijing, China, pp. 107–117.

Patrick, W.H., Jr., and C.N. Reddy. 1978. Chemical changes in rice soils. In: *Soils and rice.* Int. Rice Res. Inst., Los Baños, Philippines, pp. 361–379.

Ponnamperuma, F.N. 1972. The chemistry of submerged soils. *Adv. Agron.* 24:29–96.

Ponnamperuma, F.N. 1978. Electrochemical changes in rice soils and the growth of rice. In: *Soils and rice.* Int. Rice Res. Inst., Los Baños, Philippines, pp. 421–441.

Prasad, M., R. Prasad, and T.N. Chaudhary. 1980. Effect of planting methods on some physical properties of rice soils. *Indian J. Agron.* 25:213–218.

Prihar, S.S., K.L. Khera, and P.R. Gajri. 1976. Effect of puddling with different implements on the water expense and yield of paddy. *J. Res. Punjab Agr. Univ. (India)* 13:249–254.

Quirk, J.P. 1978. Some physico-chemical aspects of soil structural stability—a review. In: W.W. Emerson, R.D. Bond, and A.R. Dexter (eds.) *Modification of soil structure.* John Wiley & Sons, New York, pp. 3–16.

Rane, D.B., and S.B. Varade. 1972. Hydraulic conductivity as an index for evaluating the performance of different puddlers. *J. Agr. Eng.* 9:11–16.

Raymundo, M.E., E.B. Pantastico, and V.T. Manarpaac. 1962. Some physical and chemical properties of an upland soil and its associated paddy soils. *Phil. Agr.* 46:560–569.

Reddy, D.S., and S.S. Sharma. 1979. *Annual report of the Soil Physical Conditions Improvement Project, Hyderabad Centre.* A.P. Agr. Univ., Hyderabad, India.

Rengasamy, P., G. Singh, and G.S.R. Krishna Murti. 1974. Iron transformation and microaggregates in submerged soil. *Il Riso* 23:151–157.

Robinson, D.O., and J.B. Page. 1951. Soil aggregate stability. *Soil Sci. Soc. Am. Proc.* 15:25–29.

Rodriguez, M., and R. Lal. 1979. Tillage/fertility interaction in paddy rice. In: R. Lal (ed.), *Soil tillage and crop production.* Proc. Series No. 2, IITA, Ibadan, Nigeria, pp. 349–356.

Rodriguez, M.S., and R. Lal. 1985. Growth and yield of paddy rice as affected by tillage and nitrogen levels. *Soil Till. Res.* 6:163–178.

Rozanov, B.G., and I.M. Rozanova. 1962. Soils of the arid monsoon tropical zone of Burma. *Soviet Soil Sci.* 3:294–301.

Sahoo, R., A.K. Bandyopadhya, and B.B. Nanda. 1970. Effect of organic manures with and without drainage on rice yield, nutrient uptake and soil aggregation. *J. Indian Soc. Soil Sci.* 18:51–55.

Saito, M. 1985. Structure, structural stability and natural restructuring of lowland rice soils. In: *Soil physics and rice.* Int. Rice Res. Inst., Los Baños, Philippines, pp. 245–260.

Saito, M., and K. Kawaguchi. 1971a. Flocculating tendency of paddy soils. IV. Soil structure of paddy plow layers. *J. Sci. Soil Manure Jpn.* 42:95–96; also in *Soil Sci. Plant Nutr.* 18:202 (Engl. Abstr.).

Saito, M., and K. Kawaguchi. 1971b. Flocculating tendency of paddy soils. II. Structure of poorly drained paddy soils. *J. Sci. Soil Manure, Jpn.* 42:61–64; also in *Soil Sci. Plant Nutr.* 18:199 (Engl. Abstr.).

Saito, M., and K. Kawaguchi. 1971c. Flocculating tendency of paddy soils. I. Periodical changes of physical properties of paddy soils under flooded conditions. *J. Sci. Soil Manure Jpn.* 42:1–6; also in *Soil Sci. Plant Nutr.* 18:125 (Engl. Abstr.).

Sanchez, P.A. 1968. *Rice performance under puddled and granulated soil cropping systems in southeast Asia.* Cornell Univ. Press, Ithaca, New York.

Sanchez, P.A. 1973a. Puddling tropical rice soils. I. Growth and nutritional aspects. *Soil Sci.* 115:149–158.

Sanchez, P.A. 1973b. Puddling tropical rice soils. II. Effects of water losses. *Soil Sci.* 115:303–308.

Sanchez, P.A. 1976. Soil management in rice cultivation systems. In: *Properties and management of soils in tropics.* John Wiley and Sons, New York, pp. 413–477.

Savant, N.K., and S.K. De Datta. 1982. Nitrogen transformations in wetland rice soils. *Adv. Agron.* 35:241–302.

Scheltema, W. 1974. Puddling against dry plowing for lowland rice culture in Surinam. Agri. Res Rep. 823, Centre for Agricultural Publishing and Documentation, Wageningen, Netherlands, 241 pp.

Seth, A.K., C.H. Kaw, and J.M. Fua. 1971. Minimal and zero tillage techniques and postplanting weed control in rice. *Proc. 3rd Asian-Pacific Weed Conf., Kuala Lumpur, Malaysia,* 1:188–200.

Sethunathan, N., and R. Siddaramappa. 1978. Microbial degradation of pesticides in rice soils. In: *Soils and rice.* Int. Rice Res. Inst., Los Baños, Philippines, pp. 479–497.

Sharma, P.K., and S.K. De Datta. 1985. Puddling influence on soil, rice development, and yield. *Soil Sci. Soc. Am. J.* 49:1451–1457.

Shoji, S., K. Watanabe, S. Fukazawa, F. Higuchi, S. Saito, and S. Watanabe. 1974. Influence of percolation on the growth and grain yields of rice plant and physico-chemical properties of paddy soil. *J. Sci. Soil Manure, Jpn.* 45:441–446; also in *Soil Sci. Plant Nutr.* 21:197 (Engl. Abstr.).

Singh, B., and B.S. Sandhu. 1979. Effect of irrigation, mulch and crop canopy on soil temperature in forage maize. *J. Indian Soc. Soil Sci.* 27:225–235.

Singh, N.T., A.S. Josan, and J.P. Gupta. 1977. Complementary effect of soil puddling, submergence, and organic matter on rice production. *Agron. J.* 69:882–884.

Singh, N.T., M.S. Patel, R. Singh, and A.C. Vig. 1980. Effect of soil compaction and water use efficiency of rice in a highly permeable soil. *Agron. J.* 72:499–502.

Sinha, M.P. 1964. A study of the measurement of puddling and comparative performance of different implements for puddling in rice cultivation. *Proc. 3rd Meeting of the International Rice Commission Working Party on the Agricultural Engineering Aspects of Rice Production, Storage and Processing,* Manila, Philippines, Mar. 7-14, 1964.

Sinha, B.K., S.R. Singh, and R.G. Sharma. 1981. Note on puddling effects on water flow through soils. *Indian J. Agr. Sci.* 51:460–461.

Springer, M.E. 1950. Reconnaissance soil survey of Japan. Shikoku. Supreme Commander Allied Powers Natural Resources Section Report 110, 60 pp.

Sugimoto, K. 1969. Studies on plant-water relationship of paddy in Muda river irrigation project area of west Malaysia. Dept. Agr., Malaysia, 145 pp.

Swanson, C.L.W., R.M. Hanna, and H.C. de Roo. 1955. Effects of excessive cultivation and puddling on conditioner treated soils in the laboratory. *Soil Sci.* 79:15–24.

Syarifuddin, A.K. 1982. Tillage practices and methods of seeding upland crops after wetland rice. In: *Report of a Workshop on Cropping Systems Research in Asia.* Int. Rice Res. Inst., Los Baños, Philippines, pp. 33–41.

Syarifuddin, A.K., and H.G. Zandstra. 1978. Growing rainfed corn and soybeans after puddled flooded rice: I. Soil physical conditions and management. IRRI Saturday seminar, 4 Nov 1978, International Rice Research Institute, Los Baños, Philippines (mimeo).

Tadano, T., and S. Yoshida. 1978. Chemical changes in submerged soils and their effect on rice growth. In: *Soils and rice.* Int. Rice Res. Inst., Los Baños, Philippines, pp. 399–420.

Takai, Y., T. Koyama, and T. Kamura. 1969. Microbial metabolism in reduction process of paddy soils. V. Effects of rice plant roots and percolating water on the reduction process of flooded paddy soil in pot. *J. Sci. Soil Manure Jpn.* 40:15–19; also in *Soil Sci. Plant Nutr.* 15:135 (Engl. Abstr.).

Takai, Y., H. Wada, H. Kagawa, and K. Kobo. 1974. Microbial mechanism of effects of water percolation in Eh, iron, and nitrogen transformation in the submerged paddy soils. *Soil Sci. Plant Nutr.* 20:33–45.

Taneja, M.L., and S. Patnaik. 1962. A technique for determining the degree and depth of soil puddle. *Rice News Teller* 10:27–28.

Taylor, H.M. 1972. Effect of drying on water retention of a puddled soil. *Soil Sci. Soc. Am. Proc.* 36:972–973.

Taylor, H.M., A.C. Mathers, and B.F. Lotspeich. 1964. Pans in the Southern Great Plain soils. I. Why roots restricting pans occur. *Agron J.* 56:328–332.

Taylor, H.M., G.M. Roberson, and J.J. Parker, Jr. 1967. Cotton seedling taproot elongation as affected by soil strength changes induced by slurrying and water extraction. *Soil Sci. Soc. Am. Proc.* 31:700–704.

Tripathi, R.P., and B.P. Ghildyal. 1974. Thermal properties of fine textured Mollisols under varying moisture conditions. *J. Indian Soc. Soil Sci.* 22:103–108.

Tyagi, N.K., N. Acharaya, and P.C. Mohanty. 1975. Effect of puddling instruments on percolation losses and water-use efficiency in rice field. *Indian J. Agr. Sci.* 45:132–135.

Unger, P.W. 1979. Effect of deep tillage and profile modification on soil properties, root growth and crop yields in United States and Canada. *Geoderma* 22:275–295.

Utomo, W.H., and A.R. Dexter. 1981. Soil friability. *J. Soil Sci.* 32:203–213.

Vamadevan, V.K. 1971. Temperature regimes under different water depths and their effects on the growth and yield of rice. *Il Riso* 20:21–29.

van Breemen, N. 1976. *Genesis and solution chemistry of arid sulfate soils in Thailand.* Centre for Agricultural Publishing and Documentation, Wageningen, Netherlands.

Varade, S.B., and B.P. Ghildyal. 1967. Mechanical impedance and growth of paddy in artificially compacted lateritic sandy loam soil *J. Indian Soc. Soil Sci.* 15:157–162.

Virmani, S.M., K.L. Sahrawat, and J.R. Burford. 1982. Physical and chemical properties of vertisols and their management. In: *Vertisols and Rice Soils of the Tropics, Symposia Papers II, 12th Int. Congr. Soil Sci. Indian Society of Soil Science,* New Delhi, India, pp. 80-93.

Warkentin, B.P. 1962. Water retention and swelling pressure of clay soils. *Canadian J. Soil Sci.* 42:189–196.

Watanabe, I. 1978. Biological nitrogen fixation in rice soils. In: *Soils and rice.* Int. Rice Res. Inst., Los Baños, Philippines, pp. 465–478.

Webb, T., and W. Armstrong. 1983. The effects of anoxia and carbohydrates on the growth and viability of rice, pea and pumpkin roots. *J. Exp. Bot.* 34:579–603.

Wickham, T.H., and V.P. Singh. 1978. Water movement through wet soils. In: *Soils and rice.* Int. Rice Res. Inst., Los Baños, Philippines, pp. 337–358.

Wien, C., R. Lal, and E.L. Pulver. 1979. Effects of transient flooding on growth and yield of some tropical crops. In: R. Lal and D.J. Greenland (eds.), *Soil physical properties and crop production in the tropics.* John Wiley & Sons, Chichester, pp. 235–248.

Yamane, I., and K. Sato. 1967. Effect of temperature on the decomposition of organic substances in flooded soil. *Soil Sci. Plant Nutr.* 13:94–100.

Yoneyama, T., and T. Yoshida. 1977. Decomposition of rice residue in tropical soils. I. Nitrogen uptake by rice plants from straw incorporated, fertilizer (Ammonium sulfate) and soil. *Soil Sci. Plant Nutr.* 23:33–40.

Yoshida, S. 1981. *Fundamentals of rice crop science.* Int. Rice Res. Inst., Los Baños, Philippines.

Yoshida, T. 1978. Microbial metabolism in rice soils. In: *Soils and rice.* Int. Rice Res. Inst., Los Baños, Philippines, pp. 455–463.

Yun-sheng, C. 1983. Drainage of paddy soils in Taihu lake region and its effects. *Soil Res. Rept. No. 81,* Institute of Soil Science, Academia Sinica, Nanjing, China.

Zandstra, H.G. 1982. Effect of soil moisture and texture on the growth of upland crops after wetland rice. In: *Report of a Workshop on the Cropping Systems Research in Asia.* Int. Rice Res. Inst., Los Baños, Philippines, pp. 43–54.

Zein El Abedine, A., M.M. Abdalla, and M.A. Ozoris. 1969. The effect of dry crusting and wet puddling treatments on the physical properties, wilting point and root growth in a sandy loam soil and a silty clay soil. *J. Soil Sci. Un. Arab Repub.* 9:239–245.

Origin, Evolution, and Classification of Paddy Soils in China

Gong Zitong*

I. Paddy Soils of China

China has 25 million hectares of paddy soils, accounting for 25% of the total cultivated land in the country. Paddy (*Oryza sativa*) output amounts to 44% of the total grain output. These soils are characterized by a long history, varying types, and great productive potential.

---

*Institute of Soil Science, Academia Sinica, Nanjing, Peoples Republic of China

Advances in Soil Science, Volume 5

A. Early History

As judged from the finding of indica rice (*Oryza sativa*) below the ruins of Hemu Ferry Crossing, paddy cultivation in China has a history of more than 7000 years. Early in the third century B.C., a vast area along the Yangtze River and the Yellow River was already commonly under the cultivation of paddy. As early as in the 1930s, Chinese pedologists paid special attention to morphological features in the formation and classification of paddy soil from the viewpoint of soil-moisture regime (Chu *et al.*, 1983). Moreover, when the formation of paddy soil was considered as a process of podzolization in the literature, Chinese soil scientists made it clear that there was a significant difference between the processes of paddy soil formation and podzolization (Hseung, 1941).

B. Variations in Soil Type

In China the area of paddy soils accounts for one-fourth of the total paddy area in the world. These soils are characterized by a great variety of types. In the northern regions, paddy soils are usually calcareous and frequently contain soluble salts, whereas in the southern regions they are acid or neutral in reaction. For the sake of rice-growing, farmers have devoted enormous efforts to remaking the natural soil conditions in building up paddy fields in various landscapes, such as terraced fields in mountainous and hilly areas, flat fields in alluvial plains, and polder fields on lake basins. Paddy soils primarily originated from zonal soils, meadow soils and bog soils. In zonal soils, there are latosols, lateritic red earth soil, red earth soil, yellow earth soil, yellowish brown earth soils, brown earth soils, and others. Meadow soils can be acid, neutral, or calcareous. Bog soils are classified into muddy bog soils, humic bog soils, peaty bog soils, and others. Under human cultivation, soils derived from all these different origins develop toward the unique morphological profile and physio chemical characteristics of paddy soil. Although paddy soil is an independent soil type, it can be readily divided into a great number of varieties based on differences in oxidation–reduction regimes, parent material, fertility status, as well as the difference in texture. In China, paddy soils are spread over nearly all provinces of the country and the richness in types is unique in the world.

C. Productive Potential

In 1980, the annual mean grain yield of rice in China was 4.1 t ha^{-1}, exceeding the average yield of the world. However, there were sharp differences between regions. For instance, for the 40,000 ha of paddy fields in the Dai Autonomous Prefecture of Yunan Province in southwestern China the yield was only around 2.2 t ha^{-1}, whereas for the 4,266 ha of paddy fields in the Suzhou Prefecture, Jiangsu Province, the yield amounted to 9.2 t ha^{-1}. There are also remarkable differences in yield within the same region. In Dai Autonomous Prefecture there are demonstration fields with rice yields of over 6.0 t ha^{-1}, while in Suzhou Prefecture there are high-yield plots with annual yields totaling 18.7 t ha^{-1} for two croppings of rice. The differences in rice yields are closely related to soil conditions. It is clear that the potential of rice production in China is still great.

II. Distribution of Paddy Soils in China

A. Nationwide Distribution

Paddy soils are found in almost all regions of the country, with 92% of them concentrated to the south of the Qinglin Mt. and the Huahe River, in the humid southeastern part of China (Figure 1). Paddy soils located north of the warm-temperature zone make up 8.2% of the total paddy soil of the country. This area is known as the northern rice-producing area, where the single-cropping paddy system prevails, mostly in river-valley plains. Paddy soils in the northern subtropical zone at lower reaches of the Yangtze River constitute 39.5% of the total paddy soils of the country. This is the middle rice-producing area, where both double cropping of rice and wheat (*Triticum vulgare*) and double-cropping of rice are practiced. They are distributed in the Lake Taihu, Lake Dongting, Lake Poyang regions and in the western Sichuan Plain as well as in plains of various tributaries of the Yangtze River. In the region to the south of the central subtropics, paddy soils account for 52.3% of total, covering those rice-producing areas of Central China, South China, and Southwest China. The predominant cropping system here is double harvest of rice and can occur at a elevation of 2700 m. In this region they are distributed centrally in the Pearl River Delta, the Han River Delta, and the west plain of Taiwan.

B. Regional Distribution

Because humans have made good use of local natural conditions, various patterns of distribution under different conditions, peculiar to the paddy soil, have been created. They include the "continuously patched" pattern in the delta of the Yangtze River, the "patched" pattern along the coast of South China, the "branch-shaped" pattern in the hilly region of Hunan and Jiangxi Provinces, the "band-shaped" pattern in the valley's between parallel mountain ridges in eastern Sichuan Province, the "spot-shaped" pattern in the flatland of Yunan-Guizhou Plateu, and the "point-shaped" pattern on the fan margin of the inland basin of Xinjian Province (Figure 2).

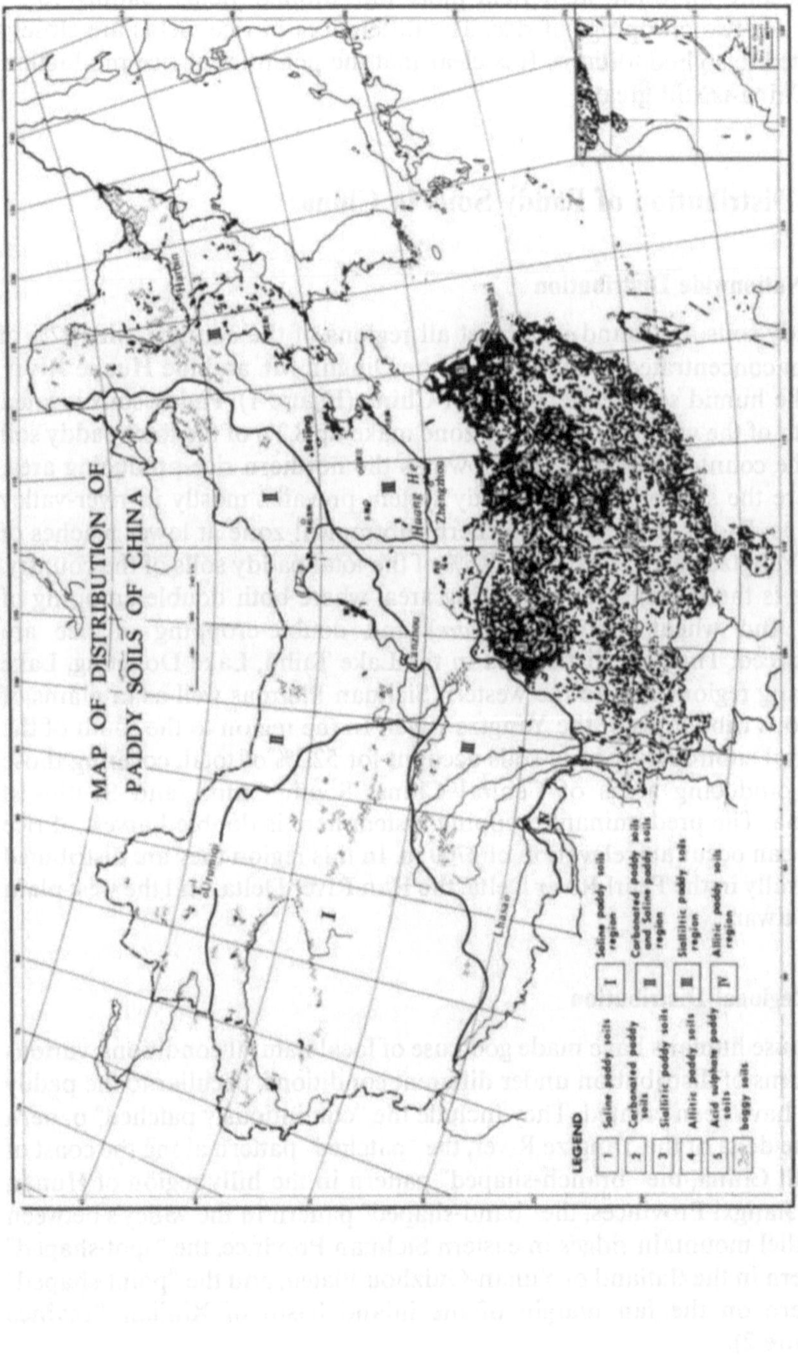

Figure 1. Distribution of paddy soil in China.

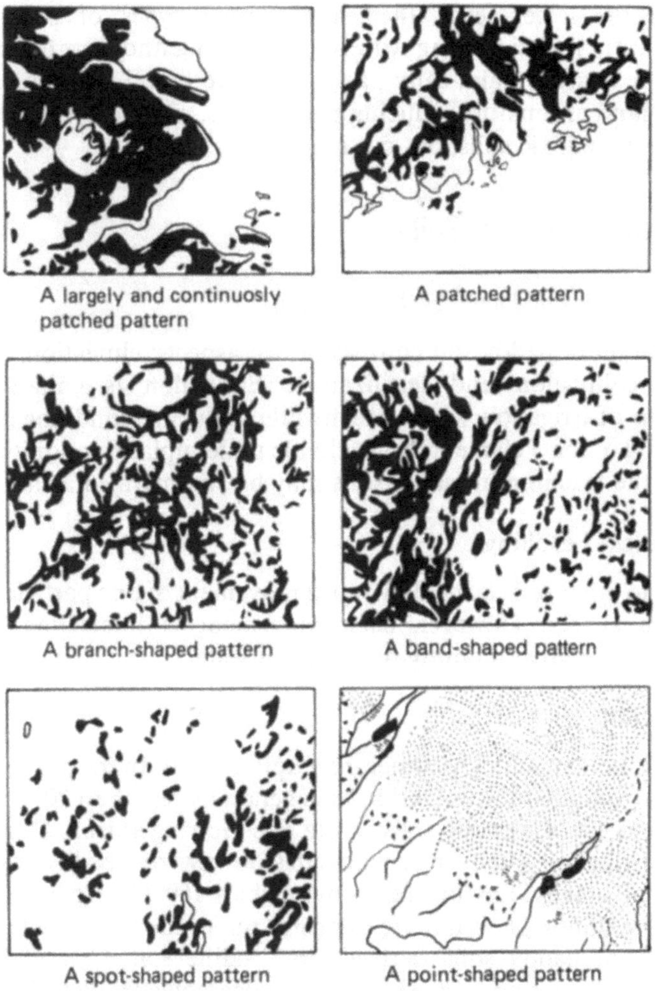

Figure 2. Characteristics of different regions of paddy soils in China.

C. Microregional Distribution

Each of the above-mentioned regions shows association features of its own in the distribution of paddy soils. In mountain areas, for example, where a vast extent of zonal soils exists, paddy soils dominated by the oxidic type are commonly scattered on the terraces. In the valley, in contrast, they are mostly reducing soils with a small area of redoxic ones. In the region of river valleys, from top to bottom below the zonal soil, paddy soils change from top to bottom below the zonal soil, paddy soils change from oxidic through redoxic to reducing, whereas on the valley margins, the paddy soils are frequently of the bleached type because of the leaching of clay and iron–manganese. In the plains area, from the higher to the lower level, there are various types of paddy soil, predominantly redoxic, along with some reducing and very little oxidic soil. On the coastal plain, salty paddy soils are found. On the dishlike low-lying land are mainly reducing paddy soils, while on its upper part there may be redoxic soils, but not oxidic ones.

III. Genesis of Paddy Soils

A. Characteristics of Paddy Soil Formation

The formation of paddy soil consists of two aspects: eluviation, and the effect of cultivation and fertilization. Eluviation is enhanced by percolation of irrigation water. Meanwhile, reducing conditions enable the leaching of some elements unleachable under upland conditions. The nature of the leaching in paddy soils may be summarized as follows.

1. Mechanical Leaching

Common to other soils is the downward movement of fine particles along large pore spaces. However, such movement is intensified by the percolation of a great quantity of irrigation water.

2. Dissolution

Paddy soils hold much more water than upland soils. In the plowed layer, the moisture content for upland soils is generally less than 20%, while it can be up to 45% for paddy soils in the rice-growing season. During the period of irrigation, such water percolates downward steadily under the influence of gravity, amounting to more than 1000 mm per year. The effect of dissolution is even more intensive under the hot climate during irrigation. For paddy soils derived from saline soil, Cl and Na are subjected to dissolution first, while in the calcareous paddy soil, calcium and magnesium carbonates are the main elements subject to dissolution and leaching. In neutral paddy soils, exchangeable bases are the principal elements subjected to leaching. In acid paddy soils, the chief change is the redistribution of exchangeable bases in the soil profile along with the manures applied in the plowed horizon. The leaching of potassium is quite remarkable, increasing sharply soon after fertilization and then decreasing quickly afterwards. Calcium and magnesium accumulate in the surface layer at first and then migrate downward gradually. Usually the illuvial horizon of magnesium is deeper than that of calcium.

3. Reductive Eluviation

Some elements with variable valency have a very low solubility in their high-valency state. However, their activity increases to a great extent in

Table 1. Distribution of some elements in paddy soils

Soil type	*Horizon*	*Fe (%)*	*Mn (ppm)*	*Cr (ppm)*	*Co (ppm)*	*Zn (ppm)*	*Ni (ppm)*	*Li (ppm)*
Saline	A	1.41	245	27.9	6.9	31.0	16.3	14.9
	B	3.43	258	74.8	9.2	68.5	32.5	54.1
Allitic	A	0.55	26	12.0	0.9	15.0	3.5	13.4
	B	2.18	257	41.0	9.5	44.5	27.3	21.3

Cited from Yang, in: *Proceedings of Symposium on Paddy Soil,* pp. 754–758, by permission of Science Press, Beijing.

the reduced state. Iron, manganese, cobalt, and nickel (Table 1) are four elements with variable valency usually found in paddy soil. Among them, iron and especially manganese are most active in eluviation under reduced conditions. Their activity is more remarkable in acid soil conditions than under neutral conditions. The amount of ferrous and manganous ions formed under reducing conditions may sometimes even exceed that of base ions. Following the rice-planting period, the illuviation of iron and manganese increases in quantity, and the depth of illuviation is lowered gradually. The illuvial horizon of manganese usually lies beneath that of iron. Under the influence of long-term reductive eluviation, a low-iron-containing "whitish soil" horizion can be formed in a light-textured soil.

4. Chelation

The difference between chelation and reduction lies primarily in that the former, although unable to change the valence state of iron and manganese, can render these elements to transfer from the solid phase to the liquid phase owing to the affinity of organic chelating agents. The ease of chelation of iron and manganese in different horizons is related to the status of their oxides. It is easier for iron to be chelated from the cultivated layer than from the illuvial horizon, with the subsoil of red earth the most difficult. It is commonly believed that chelation facilitates the leaching of iron and manganese in addition to reduction.

5. Ferrolysis

Ferrolysis is a process in which the exchangeable ferrous iron formed under reducing conditions is desorbed after drainage and the exchange sites are then occupied by hydrogen, which is in turn transformed into exchangeable aluminum. This process causes acidification of the soil and destruction of clay minerals, resulting in the migration of aluminum. It has been assumed that the acidification and the increase in the SiO_2/Al_2O_3 ratio in the colloid fraction of some neutral paddy soils of China is related to this process.

Yu *et al.* (1983) explained the effects of reduction, dissolution, and chelation in the formation of paddy soils. Brinkman (1970) pointed out the importance of ferrolysis. It is quite difficult to distinguish the above-stated five mechanisms in the genesis of paddy soil, because very often in the migration of one element several mechanisms are involved. Irrigation water hastens the effects of mechanical leaching and dissolution, and chelation may superimpose on the effect of reduction, while ferrolysis is a chemical reaction happening after reduction. To sum up, it may be said that the effects mentioned above all relate to submergic reduction. Hence all of them are called reductive eluviation.

B. Essence of the Formation of Paddy Soil

In order to understand the essence of the formation processes of paddy soil, a comparsion was made between irrigation water and percolating water in red earths and paddy soils (Table 2). Based on the coefficient of aqueous migration, the order of migration of some elements in red earths is Ca > Na > Mg > K > SiO_2 > P, Mn, Fe, Al, and that of the paddy soils is Ca > Mn > Mg > Na > P > K > Fe > SiO_2 > Al, indicating that in the formation processes of paddy soil there is an increase in the activity of Fe, Mn, K, Ca, Mg, and P. Results from the averages of five measurements in seven soil column experiments showed similar trends (Zhang, 1983), namely the concentrations of K^+, Ca^{2+}, Mg^{2+}, and PO_4^{3-} were positively correlated with those of Fe^{2+} and Mn^{2+}. Citing the results of an experiment made by Kawaguchi and Kyuma (1977), Yamane (1979) points out an increase in Ca concentration in soil solution due to the replacement of Ca absorbed on soil colloids by Fe after reduction. According to Goswami *et al.* (1979), under submerged conditions, an ion exchange can take place as a result of the increase in Fe and Mn, thus raising the concentration of potassium in soil solution. Brinkman (1970) believed that Fe could replace various kinds of base ions from the surface of soil colloids. Ponnamperuma (1980) and Goswami *et al.* (1979) are of the opinion that the reduction of ferric phosphate is the chief cause of the increase in the availability of soil phosphorus. Thus, the formation of paddy soil can be visualized as follows: under submergic reducing conditions, ferrous ions increase in amount and are adsorbed on the surface of soil colloid replacing some base ions, such as Ca, Mg, K, and Na, and rendering them into solution. Following the reduction of free iron oxides, phosphorus is liberated from the occluded form, thus raising its availablity. For the same reason, some silicon combined with iron in the form of complex salts is liberated too. However, as drying and wetting alternate the soil becomes acid and some colloids are destroyed by ferrolysis, resulting in increased exchangeable magnesium and free aluminum. On the other hand, irrigation and fertilization can bring minerals and organic substances to the soil to replenish those elements absorbed by plants and lost by leaching. If there were only reductive eluviation, the soil would eventually become extremely poor. Hence, the essence of the formation of paddy soil consists of two components: the reductive eluviation, and the wet cultivation and fertilization.

C. Stages in the Formation of Paddy Soils

In the formation of paddy soils the reductive eluviation component is bound to be more reflective of the characteristics of stages in the formation of paddy soils. Yu (1983) has classified the leaching into three stages: namely strong, moderate, and weak. From the viewpoint of the

Table 2. Chemical composition of percolation water (Jiangxi)

Water	Composition (mmol L^{-1})						
	Ca	*Mg*	*K*	*Na*	*Al*	*SiO*$_2$	*P*
Irrigation water	0.28	0.053	0.038	0.026	—	0.025	0.0001
Percolation water of red earth	0.11	0.049	0.015	0.22	Tr	0.27	0.0008
Percolation water of paddy soil	1.48	0.79	0.079	0.17	0.01	0.21	0.015

geochemical developmental stages of the soil, the leaching process can be divided into four stages: saline, calcareous, neutral (siallitic), and acid (allitic). In nature, although not very often, there are cases in which all four of these stages have occurred since the beginning of rice-planting. For saline paddy soils, the leaching process begins only from saline soils, while the calcareous paddy soil can be derived either through the desalination of saline soils or directly from calcareous soils. Neutral paddy soils are derived either from the former two types or directly on siallitic soils. Most of the allitic acid paddy soils are formed directly on allitic soils. Through long-term liming, allitic paddy soils can be changed to calcareous paddy soils.

IV. Genetic Horizons of Paddy Soil

A. Principal Genetic Horizons of Paddy Soil

In the formation of paddy soil, reductive eluviation plays a prominent role in the differentiation of horizons, especially in the formation of rusty spots and streaks, gray coatings, and gley spots. Along with the advance in soil-forming processes of paddy soil, some characteristic genetic horizons, known as plowed horizons, plow pans, and illuvial horizons, are formed. The illuvial horizon is the diagnostic horizon that can be used to distinguish paddy soil from other types of soil.

1. Plowed Horizon (Ag)

This horizon is formed under the influences of frequent cultivation and seasonal submergence and drainage. This is also the horizon for most of the root system of crops. The thickness of this horizon is usually 12 to 18 cm. During the submergence season, the whole plowed horizon is submerged by irrigation water and is in a reduced state except its upper layer, which is in an oxidation state because of the diffusion of oxygen from irrigation water. Under anaerobic conditions, reducing decomposition products are partly lost in the form of gases, partly remain, and partly percolate downward along with the water. This horizon is in an oxidized state after drainage and drying. Generally this horizon can be classified into three subhorizons: (1) the uppermost (Ag1) horizon, which is a thin muddy layer composed of deposits from suspended substances; (2) the upper plowed horizon, with a fine cloddy structure and having abundant root rustings, and (3) the lower (Ag3) horizon, with a massive cloddy structure and rust coatings of different colors attached to the structural surface.

2. Plow Pan Horizon (P)

This horizon is formed by the compression of plowing. It has a compact and slightly laminate structure, a thickness of 8 to 10 cm, and a high water-retaining capacity, with a bulk density around 1.5 g cm^3. In flooding seasons, because the Eh value rises beginning from the lower part of this horizon downwards, some substances moving with the water begin to illuviate here, resulting in the formation of mottling spots and coatings. In addition, there are plow pans containing lime concretions (Pca) and gleyed plow pan (Pg).

3. Illuvial or Hydragric Horizon (B)

Some authors also call this horizon the percogenic horizon or perco-eluvial horizon (W), the former term laying stress on oxidizing illuviation and the latter on percolation eluviation. In this horizon there are more clay particles, organic substances, bases and iron, manganese, and even cobalt and nickel. Morphologically the illuviation of iron and manganese is particularly obvious. According to the degree of oxidizing illuviation, this horizon can be further divided into three subhorizons.

The Bm horizon is an illuvial horizon formed under oxidizing conditions. The ferrous and manganous ions in percolating water are oxidized by free oxygen, resulting in the formation of an iron-manganese illuvial horizon with a yellowish brown or reddish brown color. Hence this horizon can be called an oxidizing or brown illuvial horizon, corresponding to the Bm horizon of Matsui (1966).

The Bg horizon is an illuvial horizon formed by the alternation of oxidation and reduction. As the molecular oxygen is consumed by microbial activities after flooding, the ferrous ions in the downward percolating water are first absorbed on the exchange sites. They are oxized by molecular oxygen after drainage (Wada *et al.*, 1973), forming an illuvial horizon by which a typical paddy soil is characterized. Furthermore, there occurs a so-called grayization process (Mitsuchi, 1974), in which the eluviation of organic substances makes the profile dark-colored. This horizon is gray in color with mottling spots and is called the redoxic or gray illuvial horizon, corresponding to the Bmg horizon without gleyed spots (Matsui, 1966).

The Br horizon is a horizon with gley spots, formed under conditions of weak alternation of oxidation and reduction. It will not be completely oxidized even after drainage and drying. In this horizon iron and manganese differentiate only in situ and precipitate locally. Being mottled with bluish gray spots, this horizon is called a reducing or blue illuvial horizon, corresponding to the BgG horizon of Kanno (1962), the Bg horizon of Grant (1965), and the Bmg horizon with gley spots of Matsui (1966), respectively.

The plowed horizon and the illuvial horizon are the basic horizons.

Table 3. Correlation of the subdivisions of illuvial horizons suggested by various authors

<table>
<tr><th>Authors</th><th colspan="4">Illuvial horizon</th><th></th></tr>
<tr><td>Gong (1983)</td><td>Bm</td><td colspan="2">Bg</td><td>Br</td><td>Bca</td></tr>
<tr><td>Chu et al. (1938)</td><td colspan="4">W
CW</td><td>—</td></tr>
<tr><td>Tsao (1964)</td><td colspan="4">B (Bir,Bmn)</td><td>Bca</td></tr>
<tr><td>Xu (1964)</td><td>WB</td><td colspan="3">BW</td><td>—</td></tr>
<tr><td>Kanno (1962)</td><td colspan="2">Bg</td><td colspan="2">BgG</td><td>(Bca)</td></tr>
<tr><td>Matsui (1966)</td><td>Bm</td><td colspan="2">Bmg</td><td>Bmg</td><td></td></tr>
<tr><td>Grant (1965)</td><td colspan="2">B (Bir,Bmn)</td><td colspan="2">Bg</td><td></td></tr>
</table>

Accompanied with them are a number of supplementary horizons.

The L horizon is a B horizon that has been subjected to bleaching through long-term water-stagnating or constant lateral movement of water.

The G horizon is formed under the long-term influence of ground water. Horizon BG is a transition horizon between a G and a B horizon. Although these horizons are not genetic horizons of paddy soil, they can be used as one of the bases for soil classification when they exist in association with other genetic horizons.

B. Evolution of the Horizons in Paddy Soils with Different Origins

The automorphous soil undergoes differentiation in oxidation–reduction status after irrigation. Its surface layer is in a reduced state, and the subsoil in an oxidized state. Morphologically, the soil profile develops gradually from an Ag–C pattern to an Ag–P–Bm–C pattern. The water-holding capacity increases as a result of the illuviation of clay particles and iron and manganese. The depth of illuviation of iron and manganese in the profile reflects the fertility status of the soil to a certain extent.

Meadow soil is a type of soil moistened by ground water. It is in an oxidized state in the upper layer, and in a reduced state in the lower layer. During the rice-planting period, the upper layer changes to a reduced state and the lower layer remains in a reduced state, with an alternation of oxidized and reduced state in the middle stratum. Under the influence of reductive eluviation, a redistribution of iron and manganese takes place in the profile. The downward eluviation of iron and manganese is one of the characteristics of paddy soil, whereas their upward migration is a characteristic of meadow soil. When rice is planted on the meadow soil, the soddy layer transforms to a plowed horizon, and an illuvial horizon with alternations of oxidation and reduction is created, thus forming an Ag–P–Bg–G profile.

For the bog soils, there is in general a mucky horizon, showing a state of reduction in the whole profile. After rice-planting, because of drainage and the application of mud as manure, irrigation water and ground water are separated. As a result, the homogeneity in reduction in the profile is lost, and there is a differentiation into a Br horizon. The appearance of this horizon is changed into an Ag–P–Br–G profile and possibly further into an A–P–Bg–Br–C profile (Figure 3).

C. Changes in the Plowed Horizons of Paddy Soil

The plowed horizon of a paddy soil is the overall reflection of the fertility status of the soil and is also the basis for the classification of soil species. Traditionally, such horizons can be classified into a number of types according to their tilth characteristics. Based on the experience of farmers and according to such indexes as the contents of sand, silt, and clay, organic matter, and total nitrogen and the structure coefficient and water-immersed bulk density, the plowed horizon is divided into tilth types by means of the mathematical method of principal components (Gong, 1983). The tilth types are soft and mellow, slush, semislush, viscous, stiff and compact, floury, dispersing, firm, and light. The water-immersed bulk density (g cm^{-3}) for these soils decreases in the following order:

dispersing	〉	floury	〉	light	〉	firm	〉	soft and mellow
0.75		0.70		0.64		0.60		0.59

semislush	〉	stiff and compact	〉	viscous	〉	slush
0.55		0.54		0.51		0.43

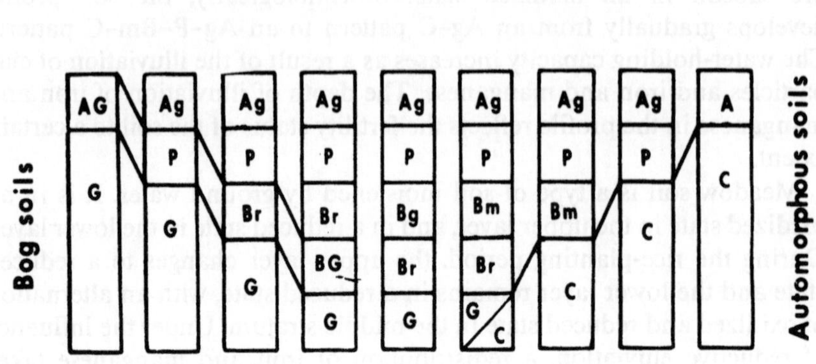

Figure 3. Profile patterns of paddy soils.

These nine tilth types are related to each other and are only different stages in the development toward the soft and mellow type of soil.

By decreasing the amounts of sand and coarse silt, which are positively correlated with the immersed bulk density, and by increasing the content of organic matter, which is negatively correlated with the immersed bulk density, the dispersing soil and floury soil can develop into soft and mellow soil through a light soil. For same reason, by increasing the amount of sand and by improving the structural status, the viscous soil and the stiff and compact soil can be changed into a soft and mellow soil. The general evolutionary patterns of these tilth types are summarized in Figure 4.

V. Classification of Paddy Soils

A. Review of Classification

The classification of paddy soils is a rather complicated problem. Different views lead to different ways of classification. Currently there are basically three ways of classification. The first, known as the soil taxonomy of the USDA (United States Department of Agriculture), is based on the viewpoint that paddy soils are merely a way of land utilization, and any changes around the form are classified only in the units below the subgroup (Mormann, 1980). The second, known as the FAO (Food and Agricultural Organization) system (1974), lays stress on the effect of water and equates the paddy soil to a gleysol. The third,

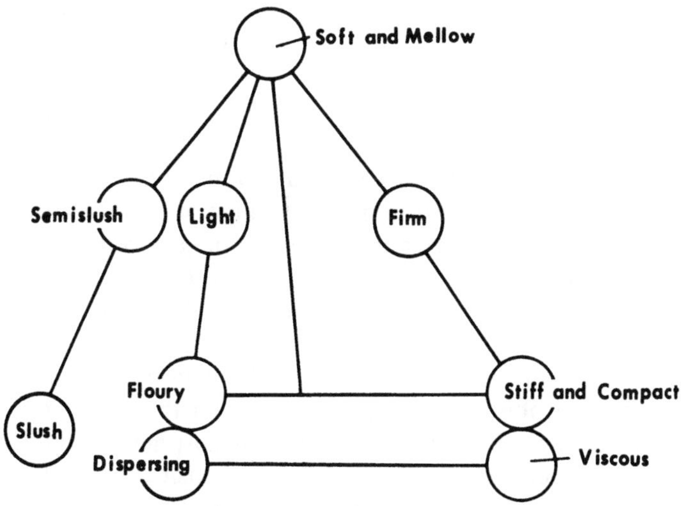

Figure 4. Patterns of plowed horizons of paddy soils

widely used in Japan, attaches importance to the special position of paddy soils in the classification. Though different in method, Kanno (1962), for example, classified paddy soils into the ground-water type, the surface-water type, and the transitional type. Kamoshita (1954) divided paddy soils into the gray lowland soil and the brown lowland soil, etc., whereas Uchiyama (from Kanno), based on the oxidation–reduction state, named paddy soils the blue reduction type, the brown oxidation type, and the greyish brown intermediate type.

In China, research on the classification of paddy soils is also quite active. On the whole, classification systems can be essentially divided into three kinds: geographical distribution, genetic factors, and genetic processes. In the classification based on geographical distribution, the great groups of paddy soils are divided in line with the bioclimatic zones, such as paddy soils in the south and those in the north. In the classification based on genetic factors, the soil types are defined largely in accordance with three factors, namely "origin," "moisture regime," and "acidity," with more emphasis on moisture regime, such as side bleaching, permeable, waterlogging, percolating, and stagnating, as proposed by Xu, *et al.* (1980). A classification based on genetic processes was suggested by Chu (1938), who, based on their studies of soil morphology along with properties, considered that paddy soils could be divided into submergic, hydromorphic, and gleyed subgroups. The present author (1980) and Yu *et al.* (1983) advanced a scheme of three subgroups, i.e., the oxidic paddy soil, the reducing paddy soil, and the redoxic paddy soil, based on a viewpoint of the oxidation–reduction process.

B. What Is a Paddy Soil?

This is a question yet to be unanimously answered. In some countries other than China, because the cropping system is not stable and the effect of rice-planting on the soil properties is still not remarkable, paddy soils are frequently considered merely as a form of land utilization. Some authors are of the opinion that only those rice-planting soils that have well-developed eluvial–illuvial horizons can be called Aquorizems (Kawaguchi and Kuyma, 1977). Currently, more and more authors are in favor of this viewpoint, because in China most of the rice-growing areas have a long-standing history and stable cropping system, and the soil is characterized by distinct horizons. Paddy soils originate from automorphous soil, meadow soil, and bog soil, but they also differ from the latter three soils.

The reductive eluviation and oxidative illuviation of paddy soil differ from the automorphous soil in that in the latter the eluviation is not as strong because the soil is only affected by atmospheric precipitation. In meadow soil, there is only an upward translocation of iron and manganese as a result of the influence of ground water, while in paddy soil, there is not only the downward leaching due to the influence of irrigation water, but also the upward translocation of some elements due to the rising of ground water. In bog soil, there is generally no downward migration of elements because of the impediment of ground water.

In the formation of paddy soils, bases, iron, and manganese are leached, while changes in silicon and aluminum are not significant. In the formation of podzolic soils, migrations take place not only in bases, iron, and manganese, but also in aluminum. Therefore, there is a distinction between the illuvial horizon of paddy soil and the B horizon of podzolic soil. In the illuvial horizon of paddy soil there are occasionally lime concretions. However, these concretions are rich in iron and manganese and thus differ remarkably from those in the illuvial horizon of pedocal (Table 4). In the illuvial horizon of paddy soil, some elements, such as iron, manganese, cobalt, and nickel, are enriched, a phenomenon that is not frequently seen in the argillic horizon. Hence the illuvial horizon of paddy soil, which differs from that of other types of soil, may be used as a diagnostic horizon.

It is proposed that only those soils which have an illuvial horizon characteristic of paddy soil should be called paddy soil. If not, they still belong to the original soil classification.

To sum up, paddy soil may be taken as an independent soil type, belonging to the suborder of the hydromorphic soil order.

C. Classification System

From the geochemical point of view and taking the diagnostic horizon as a basis, we use a classification system with five categories: namely, the great group, the subgroup, the family, the species, and the variety.

The great group is divided according to the leaching stage of paddy soil. In the first stage, known as the saline stage, substances leached out are chiefly Cl^- and $SO^=$; in the second stage, known as the carbonate state, the leached substances are mainly HCO_3^- and Ca^{++}; in the third stage, known as siallitic stage, the leached elements are Ca, Mg, K, and Na and the pH of the soil is about 7.5 (Table 5); in the fourth stage with a pH of lower than 6.5 (Table 6), known as the allitic stage, there is the migration of bases applied into the soil by fertilization. Paddy soils in the four first stages are called saline paddy soils, carbonate paddy soils, siallitic paddy soils and allitic paddy soils, respectively. What we call the saline and the carbonate paddy soils correspond to the weakly leached paddy soils proposed by Yu *et al.* (1983). The siallitic and allitic paddy soils correspond to moderately and intensively leached paddy soils, respectively.

The subgroup of a paddy soil is determined primarily on the basis of the structural pattern of the profile, which is a reflection of the genetic processes. Those patterns with an oxido-illuvial horizon (Bm) are called

Table 4. Comparison of the composition of lime concretions between paddy soil and pedocal

Kind of concretion	*No. of samples*	*Location*	*SiO_2 (%)*	*Al_2O_3 (%)*	*Fe_2O_3 (%)*	*MnO (%)*	*CaO (%)*	*MgO (%)*
In paddy soil	1	Guangxi	44.58	11.66	8.46	—	14.90	1.70
	1	Guangxi	42.58	11.05	7.84	—	14.39	1.05
	1	Guangxi	53.56	16.63	8.14	0.44	5.21	0.84
In pedocal	6	Shensi	22.21	6.21	2.04	0.04	35.59	1.56

Table 5. Chemical properties of a siallitic paddy soil (Wuxian, Jiangsu)

Depth (cm)	Horizon	pH	Composition (%)				Exchangeable bases (meq per 100 g)			
			CaO	*MgO*	K_2O	P_2O_5	*Ca*	*Mg*	*K*	*Na*
0 to 15	A	6.0	0.82	1.00	1.71	0.25	13.4	4.3	0.21	0.32
15 to 23	P	7.0	0.97	0.95	1.84	0.23	15.1	4.3	0.16	0.26
23 to 40	Bg1	7.0	0.99	1.01	1.73	0.14	14.7	5.1	0.17	0.26
40 to 60	Bg2	7.1	0.96	1.06	1.76	0.13	14.2	3.5	0.15	0.23
60 to 80	Cg	7.0	0.87	1.00	1.48	0.05	15.2	5.4	0.19	0.31

Cited from Yu, *The Physical Chemistry of Paddy Soils,* by permission of Science Press, Beijing, p. 255.

Table 6. Chemical properties of an allitic paddy soil (Taoyuan, Hunan)

Depth (cm)	Horizon	pH	Composition (%)						Exchangeable bases (meq per 100 g)			
			CaO	*MgO*	K_2O	Na_2O	P_2O_5		*Ca*	*Mg*	*K*	*Na*
0 to 12	A	5.4	0.68	0.67	1.14	0.36	0.10		8.75	0.93	0.16	0.15
20 to 35	P	6.0	0.66	0.77	1.14	0.28	0.05		8.13	1.15	0.13	0.13
45 to 60	Bg	6.8	0.64	0.74	1.25	0.35	0.04		7.25	1.20	0.09	0.12
70 to 85	B	7.0	0.84	0.69	1.24	0.42	0.02		7.13	1.25	0.12	0.11

oxidic paddy soils, those with a redoxo-illuvial horizon (Bg) are called redoxic paddy soils, those with a reducto-illuvial horizon (Br) are called reducing paddy soils, and those with horizon L are called bleached paddy soils, respectively.

The family of a paddy soil is classified according to the influence of parent materials. The classification of species is in accordance with the degree of change imposed on the original plowed horizon as a result of cultivation and fertilization and, specifically, with the tilth types, in which at least the following nine types are recognized: soft and mellow, firm, light, despersing, floury, stiff and compact, viscous, semislush, and slush.

The variety of paddy soil is classified according to variations occurring within the rank of species.

VI. A Suggestion of Paddy Soil Classification in Soil Taxonomy

The paddy soil with a hydragric horizon may be considered as an anthraquic great group under a different suborder in soil taxonomy.

References

Brinkman R. 1970. Ferrolysis: A hydromorphic soil *Geoderma* 3:199–206.

Chen H.Z. 1980. Geographical distribution of paddy soils in China. In *Proceedings of a Symposium on Paddy Soil.* Edited by Institute of Soil Science Academia Sinica. Science Press, Beijing, pp. 734–740.

Chu, L.T., Y.T. Ma, T.C. Sung, and K.C. Hou. 1938. The Nomenclature of the various horizons of paddy soil. Special Soil Publication, China, Series B43, pp. 73–78.

FAO/UNESCO. 1974. Soil map of the world, 1:5,000,000, Vol. 1, Legend, Paris.

Gong, Z.T. 1980. On the genetic classification of paddy soils in China. In: *Proceedings of a Symposium on Paddy Soil.* Edited by Institute of Soil Science Academia Sinica. Academia Simica. Science Press, Beijing, pp. 129–138.

Gong, Z.T. 1983. Pedogenesis of paddy soil and its significance in soil classification. *Soil Sci.* 135:5–10.

Goswami, N.N., 1979. Phosphorus, potassium and other macroelements. In: Soils and Rice. IRRI, Manila, Philippines, pp. 561–576.

Grant C.Y. 1965. Soil characteristics associated with wet cultivation of rice. In: *The mineral nutrition of rice plant* John Hopkins Press, Baltimore, Maryland, pp. 15–28.

Hseung Y. 1941. Some chemical properties of paddy soils. Special Soil Publication, China, Series A4, pp. 1–22.

Kanno I. 1962. A new classification system of rice soils in Japan. *Pedologist* 6:2–10.

Kamoshita, Y. 1954. Soil types in Japanese paddy rice field. *Trans. 5th Int. Congr. Soil Sci.* 4:137-139.

Kawaguchi, K. and K. Kyuma. 1977. *Paddy Soil in Tropical Asia.* The University Press of Hawaii, Honolulu.

Matsui T. 1966. A proposal on a new classification system of paddy soils in Japan. *Pedologist* 10:68–87.

Mitsuchi M. 1974. Pedogenic characteristics of paddy soil and their significance in soil classification (Engl. summary). *Bull. Nat. Inst. Agr. Sci.* B29:29–115.

Mormann, F.R. 1980. The classification of "Paddy soils" as related to soil taxonomy. In: *Proceedings of a Symposium on Paddy Soil.* Edited by Institute of Soil Science, Academia Sinica. Science Press, Beijing, pp. 139–150.

Ponnamperuma, F.N. 1980. Some aspects of physical chemistry of paddy soils. *Proceedings of a Symposium on Paddy Soil,* Edited by Institute of Soil Science, Academia Sinica. Science Press, Beijing, pp. 59–90.

Tsao, S.K. 1964. The characteristics of paddy soil derived from red earth Jiangxi Province. *Sinica Pedologica* 12:63–73. (in Chinese)

Wada, H., and S. Matsumoto. 1973. Pedogenic processes in paddy soils. *Pedologist* 17:2–15.

Xu, Q., Lu Y.C., Liu, Y.C., Zhu H.G. 1980. *The paddy soil of Taihu Region in China.* Shanghai Scientific and Technical Publishers, Shanghai. (in Chinese)

Yamane, L. 1979. Electrochemical changes in rice soils. In: *Soils & Rice.* IRRI, Manila, Philippines, pp. 392–395.

Yang, X.Y. 1980. Geochemical characteristics of some transitional and rare earth elements in paddy soils of red earth region Guangzong Province. In: *Proceedings of Symposium on Paddy Soil.* Edited by Institute of Soil Science, Academia Sinica. Science Press, Beijing, pp. 754–758.

Yu, T.R. 1983. *The physical chemistry of paddy soils.* Science Press, Beijing, pp. 234–257.

Zhang, X.Q., Xi, Q. and Y. Hseung. 1983. Influence of soil water regime on the migration of mineral elements in soil and the crop growth. *Acta pedologica Sinica* 20(4):345–360 (in Chinese).

Phosphorus Interactions with Other Nutrients and Lime in Field Cropping Systems[1]

Malcolm E. Sumner and Mart P. W. Farina*

I. Introduction

A survey of recently published textbooks and journal papers in the field of soil fertility and plant nutrition indicates that the authors seldom see fit to devote much attention to the interplay between nutrients in relation to yield enhancement in field cropping systems. This is a reflection of the widely adopted approach of varying only one or possibly two nutrients at a time in field experimentation. Without factorial designs involving a

---

[1]Contribution from the Department of Agronomy, University of Georgia, Athens, GA 30602 and the Grain Crops Research Institute, Department of Agriculture, Pietermaritzburg, South Africa
*Professor and Research Leader, respectively

Advances in Soil Science, Volume 5

number of nutrient factors at varying levels, serious study of interactions is not possible. Until field research in soil fertility pays greater attention to the multifactor approach, there will be little prospect of verifying the large number of nutrient interactions previously studied under greenhouse or solution culture conditions.

Two or more growth factors are said to interact when their influence individually is modified by the presence of one or more of the others. When factors in combination result in a growth response that is greater than the sum of their individual effects the interaction is positive, and when the combined effects are less the interaction is negative (Figure 1.). In the former case the factors are synergistic, whereas in the latter they are antagonistic. Additivity indicates the absence of interaction. Because of the complex interdependence of growth factors in biological systems numerous interactive effects have been identified. Undoubtedly, more remain to be discovered.

The recognition of the importance of nutrient balance in crop production is an indirect reflection of the contribution of interactions to yield. The highest yields are obtained where nutrient and other growth factors are in a favorable state of balance. As one moves away from this state of balance, nutrient antagonisms are reflected in reduced yields.

Despite the fact that interactions are seen by many to be an important key to future advances in farm profitability (Wagner, 1979, 1981), there are relatively few reports in the literature of their study under field conditions. There are still fewer reports indicating where knowledge gained about interactive effects has been used to advance field cropping technology and there can be little doubt that this is an aspect of soil fertility research that warrants closer investigation. A major difficulty in

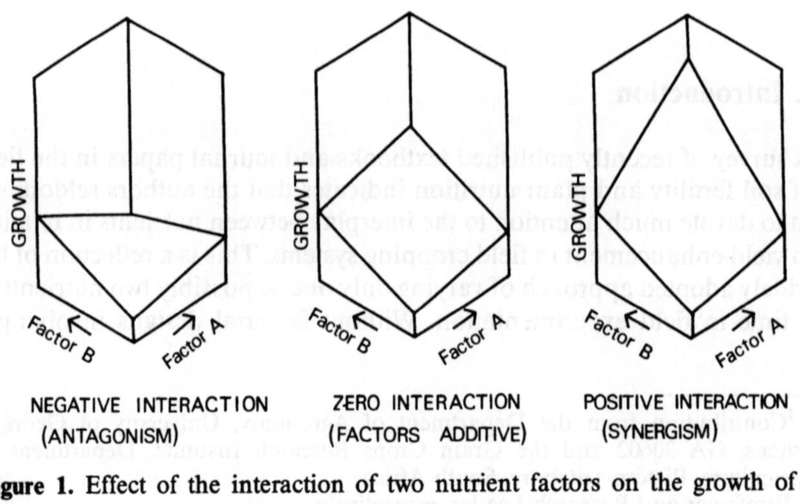

Figure 1. Effect of the interaction of two nutrient factors on the growth of a crop.

this regard is the problem of delineating the effects of uncontrolled variables, such as temperature and rainfall, unless the experiments are of longer term than those commonly reported. Then, too, there are the experimental design and logistical difficulties associated with the investigation of multiple interaction effects.

Most complete factorial experiments involving more than four factors require a large number of plots covering considerable areas of land. This problem cannot be solved by resorting to partial factorials because of the severe loss of information involved in fractional replications. Assuming that the correct experimental design has been selected, there is often difficulty in explaining even first-order interactions, particulary when supporting information, such as soil and plant analyses, is not available. Second- and higher order interactions become ever increasingly more difficult to explain. Without supporting information that reflects the effect of treatments on soil and tissue composition, little can be deduced from a statistically significant yield interaction arising from the application of two or more nutrient growth factors.

Because of the key role of P in metabolic pathways and its historically dominant position in soil fertility research, it is perhaps not surprising that recorded interactions between P and other nutrients and lime abound and a comprehensive listing of P interactions, in both the soil and the plant, and their postulated mechanisms has recently been published (Adams, 1980). The vast majority of P interactions, however, have been recorded in laboratory and greenhouse studies and their field significance is frequently debatable. In the space available Adams (1980) could devote little attention to the agronomic implications of P interactions and it is the intention here to endeavor to supplement his review with an examination of those aspects more directly applicable to field cropping. An attempt will be made to illustrate some of the problems affecting the identification and interpretation of P interactions under field conditions and the manner in which such information may assist in the optimization of field cropping techniques. Adams' (1980) review, which covers the literature up to 1976, is recommended for a more complete understanding of the possible mechanisms involved and for necessary information on the breadth and scope of identified interactions that may in time also be shown to have importance in field cropping. It should be clearly understood that the present contribution does not purport to be an exhaustive exploration of the literature but rather an attempt to look at interactions of P and other nutrients from new perspectives.

Before the various interactions between P and other nutrients are discussed, it is important to develop a model against which reported results in the literature can be evaluated. The published literature in this field often reflects contradictory results that are often not readily reconcilable. Thereafter, the various interactions are treated in separate sections that are composed of a brief synopsis of the findings of Adams (1980), together with a discussion of the pertinence of the interactions to field cropping systems. Wherever possible use is made of the model in attempting to bring clarity and understanding to the discussion.

II. Interpretive Model

The model to be used as a basis for interpreting and attempting to understand interactions is based on Liebig's law of the minimum, which states that "even if all but one of the essential elements be present, the absence of that one constituent will render the crop barren," and Mitscherlich's law of physiological relationships, which states that "yield can be increased by each single growth factor even when it is not present in the minimum so long as it is not present in the optimum." Essentially these two laws say the same thing except that Mitscherlich's law is less rigorous in its conditions and does take account of the fact that yield responses are obtained to growth factors that are not primarily limiting. For example, yield responses to fertilizers can be obtained in both wet and dry seasons despite the fact that moisture may be the most limiting factor in the latter case.

The model to be used is illustrated in a generalized form in Figure 2

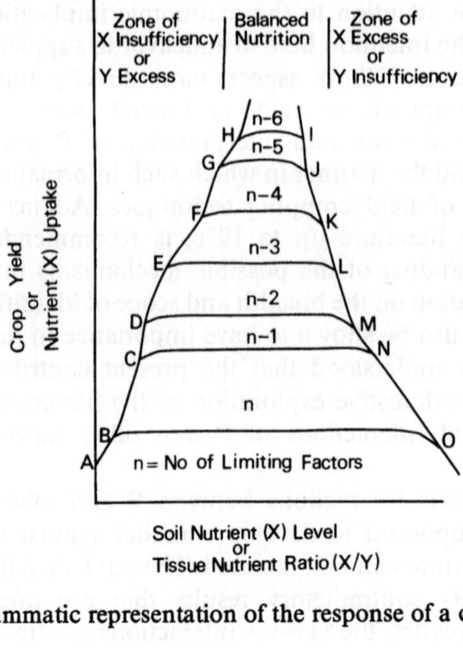

Figure 2. Diagrammatic representation of the response of a crop to a number of limiting factors.

and is applicable no matter which parameters are plotted on the ordinate and abscissa. For the sake of the discussion to follow, crop yield and tissue nutrient ratio are selected.

At point A, yield is low as a result of an insufficiency of nutrient (X) or an excess of nutrient (Y). As this state of imbalance is corrected, perhaps by increasing the soil level of X, yield increases to point B at which one or more of the many limiting factors (n) determines the yield plateau. Further increase in the level of X (or decrease in Y) will result in no further yield increase, with yield remaining constant along the line BO. At point 0, a state of reverse imbalance severe enough to reduce yield has been induced. If the level of the most limiting of the n factors is raised to a nonlimiting status, the yield would increase as a result of further increases in X to point C at which it would level off as a result of the $n - 1$ factors still limiting yield. After further substantial increases in X, a point N is reached where imbalance would again result in a yield decrease. The situations which are induced at points N and O would correspond to negative interactions or antagonisms, whereas the increase in yield from B to C as a result of increasing both the most limiting of the n factors as well as X could reflect a synergism or positive interaction. The yield increase from C to D results from the elimination of one of the $n - 1$ factors limiting growth with very little yield response to X. Such a case could reflect a situation of no interaction or additivity. As more and more of the n limiting factors are corrected, yield increases in response to addition of X. It would be generally true to state that the higher the yield the more sensitive the crop becomes to imbalances between the various growth factors. The sensitivity to factors X and Y would be measured by the length of the horizontal lines at various yield levels. A model such as that presented in Figure 2, despite its simplistic nature, is useful in attempting to explain various interactions reported in the literature.

Before the individual interactions between P and other nutrients are discussed it is perhaps opportune to make a point often overlooked by research workers in soil fertility. All too often one finds numerous reports in the literature of the relationship between yield and the quantities of nutrients applied to the soil without any presentation of meaningful soil or plant analyses. Such an approach, which is aptly named "spread and measure," contributes nothing to our understanding of the processes involved in the yield responses measured. Indeed if anything, this approach merely adds worthless pages to an already voluminous literature and, into the bargain, probably leads to confusion.

When a nutrient treatment is applied to a soil, the soil reacts to the treatment in a way dependent on the unique characteristics of the particular soil. For example, the result of adding 40 kg P per hectare to an Oxisol, on the one hand, and to an Entisol, on the other, would be very different in terms of the amount of P that would be found in the "available" pool. Thus the plant root does not respond to the treatment per se but to the soil's response to the treatment, which is usually assayed by some form of soil analysis. Without soil analysis one is uncertain of how much of the added P was being offered to the plant. Furthermore the plant shows a response to the soil's response to the treatment and it is therefore necessary to analyze the plant in order to determine how it reacted. Without plant analysis, one cannot determine whether or not the plant has reacted in the expected manner and whether its composition is in a state of nutrient balance. Without the supplementary tools of soil and plant analyses, it is doubtful whether there is any merit in measuring interactions of applied levels of two or more factors of yield.

III. Phosphorus × Nitrogen

Because of the dominant role of N and P as fertilizer nutrients in most cropping systems, P–N interactions are probably economically the most important of all interactions involving P. Considering the large volume of literature dealing with P–N interactions in greenhouse studies, however, there are surprisingly few reports emanating from field trials. Moreover, many such reports do little more than report on the existence or otherwise of an interactive effect and the practical consequences are rarely explored. In large part this is because in the agriculturally developed areas of the world, both N and P fertilization is liberal and interactive effects are, indeed, probably of lesser importance (Engelstad and Allen, 1971). However, the serious paucity of suitably designed and sited field experiments is undeniably also a contributing factor. Too little is known about the effects on final yield.

Typically, P–N interactions result in yield responses similar to those shown in Table 1. In this instance, where differential rates of limestone ammonium nitrate and superphosphate were incorporated prior to planting (Farina and Channon, 1980), response to both N and P was small at "deficient" levels of the other nutrient, increased markedly at

Table 1. Interactive effects of N and P on maize (*Zea mays* L.) grain yields (Mg ha^{-1}) (Farina, unpublished data)

$NaHCO_3$ soil P ($\mu g\ ml^{-1}$)	*N applied[a] ($kg\ ha^{-1}$)*			
	0	*60*	*120*	*180*
3	2.31	2.22	3.11	3.41
16	4.74	6.68	7.91	7.84
27	4.06	7.14	9.12	9.74
46	4.54	8.17	9.42	9.96

[a]LSD (0.05) = 0.82.

higher rates, and subsequently decreased as current yield limits or levels of "sufficiency" were approached. This pattern of behavior is in accord with the model presented in Figure 2. Under field conditions, statistical identification of such interactions may be difficult or impossible if the levels of insufficiency are only marginal or, as is the case with many leguminous crops, if N is supplied by rhizobial fixation. That such interactions should be largely ignored in systems of highly developed agriculture is perhaps understandable. However, in the less developed areas of the world, their significance cannot be overlooked. The effects on yield are obviously large and very meaningful.

Numerous workers have demonstrated that P–N interactive effects on yield are primarily attributable to N-induced increases in P absorption by the plant (Terman *et al.,* 1977). The mechanisms involved are not well understood, but a number of both soil- and plant-related mechanisms have been proposed (Grunes, 1959; Adams, 1980). The magnitude of such increases in plant P uptake may be large and it is interesting to note that, in the instance cited here (Table 1), leaf P content at flowering (Table 2) was closely related to the final yields obtained. It is particularly noteworthy that increased P uptake has also been noted on high P soils (Terman and Noggle, 1973) and in field trials where there have been no yield responses to applied P (Farina, unpublished data). In the latter case, N applications resulted in a 50% increase in leaf P content at tasseling, even though the mean yield of dryland maize exceeded 11 Mg ha^{-1}. This observation can also be accomodated by the model. The lack of response is undoubtedly the result of one of the *n* uncontrolled factors now becoming limiting. It is evident (Table 2) that the P content of plants grown without added N remained marginal even at the highest level of soil P but increased well into the "sufficiency" range at high N rates. Since dilution effects were greatest at high rates of applied N (Table 1), it would appear that N played a critical role in insuring plant utilization of available P.

It is a commonly held view that increased growth requires more of both

Table 2. Interactive effects of N and P on the P content (%) of maize leaves at flowering (Farina, unpublished data)

$NaHCO_3$ soil P ($\mu g\ ml^{-1}$)	*N applied[a] ($kg\ ha^{-1}$)*			
	0	*60*	*120*	*180*
3	0.13	0.12	0.15	0.15
16	0.19	0.24	0.25	0.26
27	0.20	0.26	0.32	0.32
46	0.24	0.32	0.34	0.35

[a]LSD (0.05) = 0.03.

Table 3. Interactive effects of N and P on the N content (%) of maize leaves at flowering (Farina, unpublished data)

$NaHCO_3$ soil P ($\mu g\ ml^{-1}$)	*N applied[a] ($kg\ ha^{-1}$)*			
	0	*60*	*120*	*180*
3	2.17	2.35	2.70	2.89
16	2.19	2.73	2.90	3.21
27	2.09	2.64	3.17	3.24
46	2.06	2.88	3.08	3.27

[a]LSD (0.05) = 0.16.

N and P, the inference being that mutually synergistic effects result in growth stimulation and enhanced uptake of both elements. This may be an oversimplification, however. In the trial from which these data were taken, the interactive effect is, indeed, also evident in plant N content (Table 3). However, it is far less marked. There are fewer reports indicating a strong stimulatory effect of P on N content and these results are, thus, in accord with many of the mechanistic studies reported in the literature (Grunes, 1959; Adams, 1980).

In view of the fact that N has a synergistic effect on the P content of the maize leaves in this investigation and vice versa, as well as having positive interactive effects on yield, the correct way of presenting the data in terms of testing the model would be to not use the ratio of N to P in the tissue but the product N $\times$ P. The use of the N/P ratio would imply that N and P applications are antagonistic in their tissue contents, which is not the case here. This product (Table 4) can be considered as a ratio of P to the reciprocal of N (P/1/N = P$\times$N). The average value for the product P$\times$N for a population of 1000 observations for maize crops yielding more than 10 Mg ha^{-1} is 1.07 $\pm$ 0.33, which is in close agreement with the values obtained here at high yields (Sumner, unpublished data). Thus in

Table 4. Interactive effects of N and P on the product of the N and P (N $\times$ P) contents (%) of maize leaves at flowering (computed from data in Tables 2 and 3)

$NaHCO_3$ soil P ($\mu g\ ml^{-1}$)	*N applied ($kg\ ha^{-1}$)*			
	0	*60*	*120*	*180*
3	0.28	0.28	0.41	0.43
16	0.42	0.66	0.73	0.83
27	0.42	0.69	1.01	1.04
46	0.49	0.92	1.05	1.14

order to use the model all we need to do is define Y as 1/N. The data plotted in this way are presented in Figure 3. When yield is plotted against the other leaf parameters the fit is not as good, with r^2 values for %P, %N, and N/P being 0.93, 0.63 and 0.57, respectively.

Of primary importance from a field cropping viewpoint is how the effects of P–N interaction can be utilized to improve production and profitability. This is particularly true of the agriculturally less developed areas of the world, where fertilizers may be in short supply, the costs are high, and soils are frequently severely P deficient. Perhaps rather obviously, field experiments designed to measure the effects of interaction provide invaluable data regarding attainable yield maxima. Multiple level and factor experimentation is equally clearly a prerequisite, as is illustrated in Table 1. It is improbable that the interactive effect would have emerged as clearly had fewer input levels been employed, and there is actually evidence that more levels might have provided further valuable information. Although the yield differences at high rates of N and P are no longer statistically significant at the 95% level of probability, the trend is unchanged. Could the P–N interaction

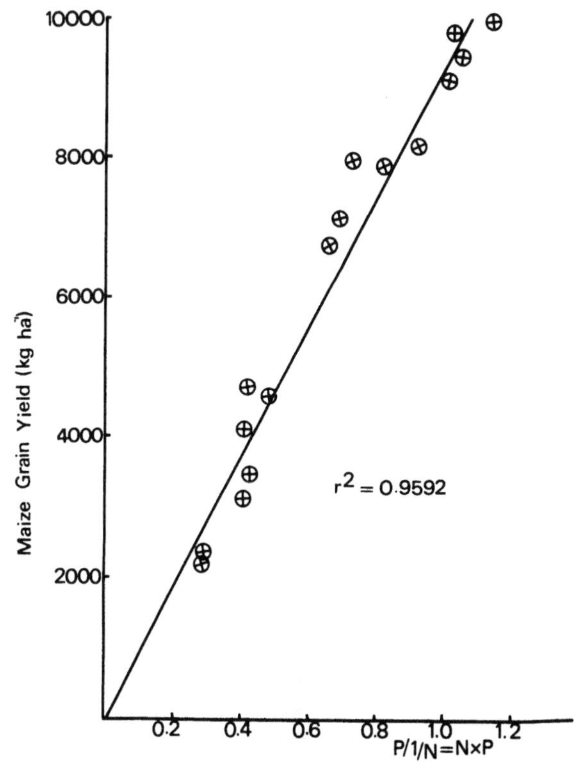

Figure 3. Relationship between maize grain yield and product of N and P contents of leaf tissue.

have been exploited further by varying other nutritional and cultural factors, which also interact with N or P? The model would indicate that the alleviation of the next most limiting factor would cause a yield increase that would probably allow the P–N interaction effect to continue until another of the *n* factors becomes limiting.

As is also evident from Table 1, yields of similar order are obtainable at widely differing N and P input levels. Since N/P cost ratios vary considerably, both in different cropping areas and over time, this interactive effect is of appreciable economic significance. Heady and Dillon (1961), Colewell (1978), and others have developed computational techniques that make possible the calculation of optimal (least cost) inputs for the attainment of specified yields. Due largely to the extensive and costly field experimentation required for the development of reliable response surface equations and because many agriculturally more advanced nations tend to produce at the high-yield levels less affected by interaction, the methodology is not as widely used as is perhaps warranted. Nevertheless, it facilitates very valuable practical exploitation of the consequences of interaction. This is conveniently illustrated by further examining the investigation from which the data discussed so far were taken (Farina and Channon, 1980).

When subject to economic analysis, the yield equation describing 3-year data indicated that least cost N and P inputs for the particular soil-bioclimatic system involved varied appreciably over the yield range experimentally attained (Table 5).

On unphosphated, virgin land the cheapest combinations of N and P vary, in terms of N and P fertilizer requirement, from 1:1 at 4 Mg ha^{-1} to 1.7:1 at 7.5 Mg ha^{-1}, with relatively large shifts occurring as the maximum yield attained is approached. Had the effects of N and P been purely additive input ratios would have been unchanged. Large shifts in N requirement relative to P clearly illustrate the practical consequences of interaction in this instance.

Table 5. Least cost N and P combinations for specified maize yields (Farina, unpublished data)

Yield (*Mg* ha^{-1})	*N rate* (*kg* ha^{-1})	$NaHCO_3$ *soil test P* (*μg* ml^{-1})
3.0	10	5
4.0	20	7
5.0	30	10
6.0	60	14
6.5	80	16
7.0	110	19
7.5	150	22

Whereas the interaction between N and P may appear relatively unambiguous in the case discussed here, this is frequently not the case in field experimentation. Other unknown or uncontrolled variables may either mask or markedly modify the nature of the interaction. Such effects are probably pertinent to all interactions with P but are less easily illustrated than they are in the case of N and P.

Soil and climatic effects on the nature of P–N interactions are particularly marked and for this reason the development of reliable response surface equations invariably requires long-term experimentation. Soil factors, such as mineralizable N supply, may take several seasons to reach a state of equilibrium (especially in newly developed areas), medium to high levels of available P may limit the range of response to such an extent that interaction is not measurable in terms of yield, and in hydromorphic soils sporadic waterlogging may drastically alter the efficiency of N utilization from one season to the next. Nitrogen mineralization poses a particularly difficult problem in that N availability cannot be reliably predicted in dryland cropping systems and legume rotations or even fallow periods may significantly alter response patterns. An example of such effects on N requirement is given in Table 6.

Although it seems obvious that long-term experimentation is unavoidable if reasonably high probability response equations are to be developed, this in itself generates further complicating factors. In poorly buffered soils, high-N plots may acidify rapidly and P–N interactions may become confounded with P–Al interactions. Similarly, such factors as nematodes, which can affect P absorption by plants (Farwell *et al.*, 1980), may significantly alter the nature of P–N response. The confounding effects of nematodes in a 3^4 N, P, K, lime experiment with maize are illustrated in Table 7. Both Al toxicity at excessively high N rates and

Table 6. The influence of previous crop, waterlogging, and fallow and N applied on yield of maize (Mg ha^{-1}) on a hydromorphic clay loam (Farina, unpublished data)

	N applied (kg ha^{-1})				
Season	*0*	*60*	*120*	*180*	*LSD (0.05)*
1973/74 (following ley)	10.25	11.51	11.56	11.62	0.42
1974/75	5.60	7.85	8.95	9.20	0.46
1975/76 (waterlogging)	2.65	4.60	6.75	7.95	0.54
1976/77	3.45	4.55	6.10	6.05	0.40
1977/78	4.45	7.00	8.35	8.55	0.30
1978/79 (fallow)					
1979/80	5.85	7.55	7.95	7.90	0.43
1980/81	4.47	7.39	8.62	8.76	0.28

Table 7. Nematode effects on maize yield (Mg ha^{-1}) response to N and P (Farina, unpublished data)

P level	*Fumigation*	*N level*[a]		
		N_0	N_1	N_2
P_0	Yes	4.02	6.07	6.32
P_1	Yes	4.15	6.97	8.23
P_2	Yes	4.27	7.17	9.34
P_0	No	3.85	5.91	4.55
P_1	No	4.32	6.15	6.37
P_2	No	4.51	7.44	7.77

[a]LSD (0.05) = 0.76.

nematodes resulted in reduced root development and disturbed P nutrition. Soil fumigation and lime both ameliorated the problem considerably, and the resulting difference in the P–N interaction due to soil fumigation is clearly evident. In unfumigated plots, the response to P was enhanced and that to N reduced.

In the light of confounding effects such as those discussed above, it is perhaps not surprising that P–N interactions observed in greenhouse and laboratory studies are frequently not as readily measured in the field and that their field occurrence is often confusingly inconsistent.

In conclusion it is perhaps pertinent to suggest that the preponderance of greenhouse over field studies reported in the literature is difficult to justify. The limitations of greenhouse studies are well known and increased field experimentation, suitably long term in nature and employing multiple level designs, may well facilitate exploitation of an interaction that is potentially of considerable practical significance. In agriculturally less developed areas of the world, it is seldom possible to fertilize for near maximum yields in the short term. Rather, the level of fertility and production has to be raised slowly over a period of time and information regarding optimum P–N input ratios is invaluable. Even in highly developed areas where current yields are considered to be near the optimum, there is, as Wagner (1979, 1981) has suggested, reason to believe that further study of interactive effects may prove profitable. It is interesting to conjecture whether or not rapid worldwide increases in N usage relative to P are not partly the result of previous deficiences in this regard. In many parts of the world there is strong evidence to suggest that this is so, decades of high P relative to N usage having been followed by the belated realization that greater N use is warranted.

IV. Phosphorus × Zinc

Recent views on this topic clearly indicate that this subject is fraught with many apparent contradictions, which have probably partially arisen as a result of the special conditions under which each investigation has been conducted (Olsen, 1972; Adams, 1980). This anomalous behavior is illustrated by the fact that some workers have found that increasing P levels reduced Zn uptake and yield as a result of an assumed P-induced Zn deficiency, while others have reported the reverse. Others again have been led to conclude that Zn uptake and yield were unaffected by P rates. As far as visual deficiency symptoms are concerned, it seems that they do not always appear concomitantly with decreases in Zn content associated with yield decreases resulting from interactions with P levels. Many workers have avoided the issue of attempting to rationalize these contradictary findings by simply postulating that other growth factors have had an influence on the P–Zn interaction. In the discussion to follow, examples of studies that illustrate this anomalous behavior are examined in the light of the model presented earlier. It is hoped that this exercise may clarify the picture of P–Zn interactions in cropping systems for the reader.

Boawn *et al.* (1954) and Ellis *et al.* (1964) reported that they failed to induce Zn deficiency in corn and beans (*Phaesolus vulgaris*) when high rates of P were applied, and in one case even despite the fact that the P content of plant tissue was more than doubled. Replotting of these data in Figure 4 to conform to the model shows that in both cases the authors had not induced a sufficient P–Zn imbalance to cause a significant yield decrease, although there was a tendency in that direction. The only significant yield responses obtained were to P. Had they applied high levels of P, the shape of the curves indicates that the probability of inducing Zn insufficiencies would have been high. These results therefore do not appear to be anomalous in the light of the model presented as they both correspond to cases in which P is relatively insufficient or Zn relatively excessive. Thus, increasing rates of P application led to yield increases.

Studies by Ganiron *et al.* (1969) with corn in the field showed that additions of P to a P-sufficient soil will not stimulate growth but may in fact significantly reduce yield as a result of an induced Zn deficiency. However, their studies indicated that this phenomenon might not always be predictable from year to year. According to Ganiron *et al.* (1969) the Zn and P contents of tissue were not related to the appearance of the condition or grain yields in either year of the study. However, computations based on their raw data for 1964 show a highly significant ($r^2 =$ 0.975) inverse relationship between P content of tissue and grain yield. Perhaps the negative correlation between these two variables, which runs contrary to the generally accepted relationship, led the original authors to believe that no relationship existed! Their data have been replotted in Figure 5 in terms of the model. The observations for 1964 now conform to a pattern predicted by the model, namely one of a relative excess of P or relative insufficiency of Zn. The treatments for a few of the points are anotated in Figure 5. When P was added, yield decreased, which was

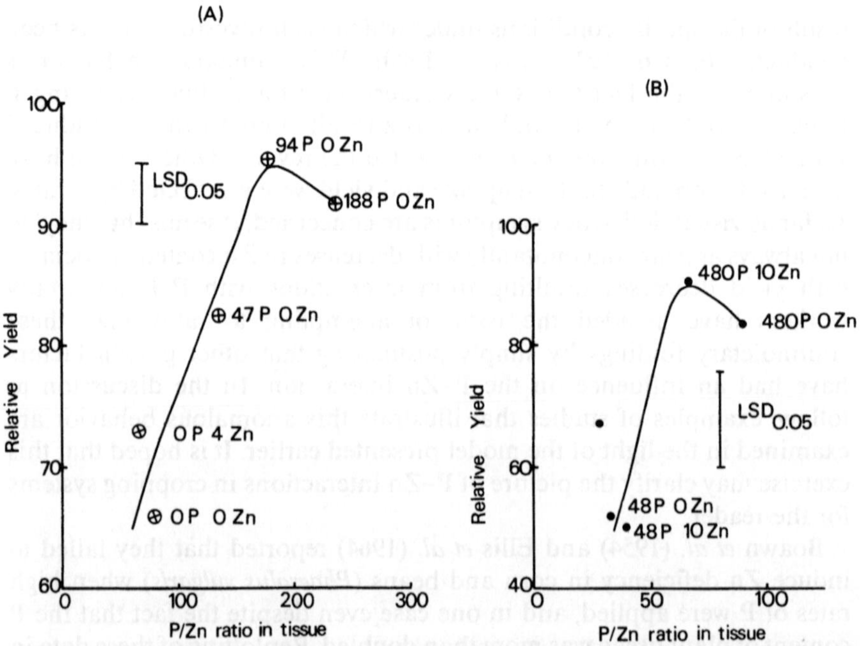

Figure 4. Relationship between relative yield (greenhouse) and the P/Zn ratio in tissue for (A) beans (*Phaesolus vulgaris*) and (B) maize. The levels of P and Zn annotated on the curves are in kg ha^{-1}. Drawn from data of Boawn *et al.* (1954) (A) and Ellis *et al.* (1964) (B).

further aggravated by the addition of a little Zn. However when adequate Zn was added yield increased dramatically. In the data for 1965, in which no significant yield response was obtained, the points are clustered and there appears to be no relation between yield and P/Zn ratio in the tissue. These data are still consistent with the model and indicate that in 1965 some factor other than P and Zn was primarily determining yield. Data of Boawn *et al.* (1954) for beans grown in the field follow the same pattern of behavior as those of Ganiron *et al.* (1969) for the 1964 season.

Examination of data from a factorial field experiment with corn in which a significant P×Zn interaction was obtained (Takkar *et al.*, 1976) shows that the model essentially predicts the behavior observed. Their data have been plotted in Figure 6. At low tissue P/Zn ratios obtained at low soil P levels, yields are low, consistent with the P deficiency symptoms. Intermediate P/Zn ratios were obtained when applications of both P and Zn were made to the soil. Under these conditions the highest yields were obtained. At high P/Zn ratios obtained with high-P and zero-Zn applications, severe Zn deficiency symptoms were observed and yield decreased accordingly.

The results obtained by many other workers when replotted in the above manner show exactly the same pattern. The crops covered by these papers and the conditions under which the experiments were conducted are listed below.

Author	Crop	Conditions
Asif *et al.* (1976)	Onions (*Allium cepa*)	Field
Spratt and Smid (1978)	Flax (*Linum usitattssium*)	Field
Singh and Singh (1979)	Wheat (*Triticum aestivum*)	Field
Stukenholtz *et al.* (1966)	Maize (*Zea mays*)	Field
Bingham and Martin (1956)	Citrus (*Citrus sinensis*)	Soil in greenhouse
Bingham *et al.* (1958)	Citrus	Soil in greenhouse
Ellis *et al.* (1964)	Beans (*Phaesolus vulgaris*)	Soil in greenhouse
Elsokkory *et al.* (1981)	Maize	Soil in greenhouse
Giordano and Mortvedt (1978)	Maize	Soil in greenhouse
Khan and Zende (1976)	Maize, wheat	Soil in greenhouse
Leece (1978)	Maize	Soil in greenhouse
Loneragan *et al.* (1979)	Clover (*Trifolium subterraneum*)	Soil in greenhouse
Safaya and Singh (1977)	Cowpeas (*Vigna sinensis*)	Soil in greenhouse
Warnock (1970)	Maize	Soil in greenhouse

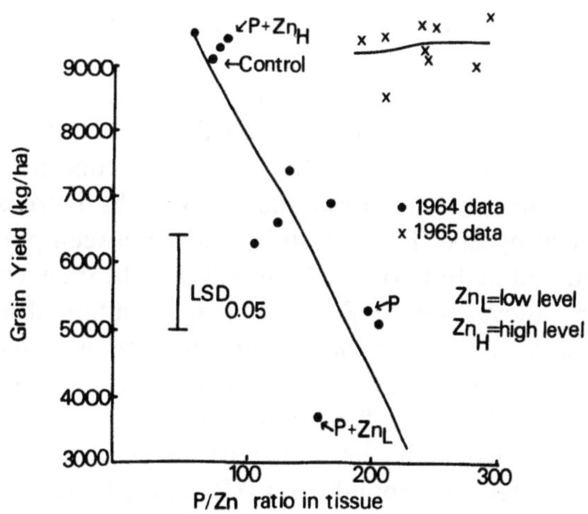

Figure 5. Relationship between maize grain yield and leaf P/Zn ratio. Drawn from data of Ganiron *et al.* (1969).

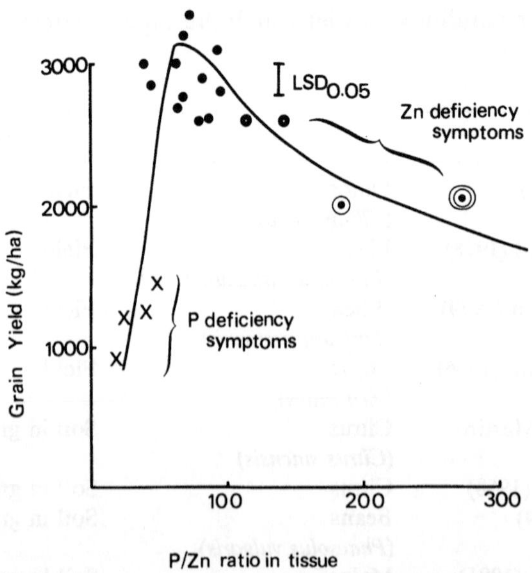

Figure 6. Relationship between maize grain yield (field) and leaf P/Zn ratio. The increasing size of circles is a measure of the intensity of Zn deficiency symptoms. Drawn from data of Takkar *et al.* (1976).

To illustrate that the model is essentially correct in its predictions concerning the other limiting factors (*n*) referred to in Figure 2, the data of Giordano and Mortvedt (1978) will be used. Unfortunately their work was conducted in the greenhouse. They studied the effect of temperature on the growth of maize and its uptake of P and Zn at two Zn levels and two P carriers. Their data are replotted in Figure 7 to show how the amelioration of a growth-limiting factor (temperature) affects the relationship between yield and the P/Zn ratio in tissue. Starting at point A on the 18°C curve, increasing the level of soil Zn results in a yield increase with a concomitant reduction in the P/Zn ratio in the tissue indicating an alleviation of the Zn deficiency condition. The least significant difference value shows that this yield increase is barely significant, indicating that the left hand portion between points G and E could be considered as horizontal. However a yield increase as great as that obtained by alleviating the Zn stress can be obtained by increasing the temperature to 27°C. Further yield increase is possible at this temperature by reducing the Zn stress until a point is reached at point F where some factor other than P/Zn and temperature becomes limiting. The LSD value shows that the FH portion of the curve could be considered horizontal. The dotted lines AB, CD, EF, etc., correspond to the lines ON, NM, etc., on the model curve (Figure 2) depending on the point on the P/Zn axis at which temperature is raised. This again is a

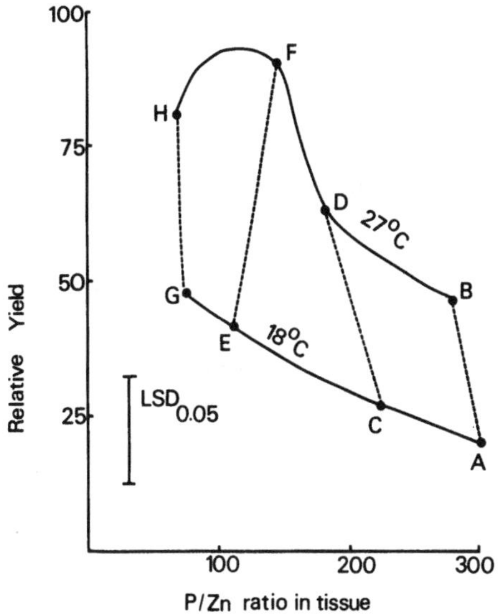

Figure 7. Relationship between relative yield and P/Zn ratio of maize leaves at two temperatures. Drawn from data of Giordano and Mortvedt (1978).

confirmation of the general applicability of the model. A similar confirmation can be obtained from the data of Singh and Singh (1979).

V. Phosphorus × Lime

Changes in soil pH brought about by liming may have profound effects on the availability of many elements absorbed by crops. In addition, most liming materials contain significant quantities of both Ca and Mg and often also of Si, and the interpretation of lime responses, particularly in field experimentation, is complicated by numerous possible confounding effects. This is particularly true of P–lime interaction studies as Ca, Mg, and silicate ions have all been shown to interact with P and so too have many of the other elements affected by pH changes (Adams, 1980). Difficult to explain second- and higher order interactions undoubtedly abound. A further complicating factor is the wide genetic variation in response to both lime and P (Foy *et al.,* 1978) and the role of rotations in many lime studies. In the light of previous discussion on P–N interactions, for example, it is clear that any increased N supply that may result from liming a previous leguminous crop could lead to the misinterpretation of apparent P–lime interactive effects in a subsequent cereal crop. It is hardly surprising, therefore, that the most striking feature of the literature dealing with lime effects on P availability is the remarkable degree of contradiction. This is especially true when earlier literature on the topic is compared with that published during the last two decades.

To keep this section within proportionate bounds, emphasis will be placed on the effects of pH per se on P availability. The possible effects of higher order interactions with P will not be examined and interactions with Mg and Si will be dealt with separately. The reader is referred to Adams (1980) for information on P–Ca interactions as little is known about their relevance under field conditions and to the recent comprehensive critical review by Haynes (1982) for further detailed information on the effects of pH on soil P chemistry. Because of the paucity of literature derived from field experimentation, extensive reference has had to be made to data obtained in greenhouse and laboratory studies.

Liming has been reported to increase (Truog, 1953; Soltanpour *et al.*, 1974; Ryan and Smillie, 1975; Jaakkola *et al.*, 1977; Parfitt, 1977; Smythe and Sanchez, 1980), decrease (Amarasiri and Olsen, 1973; Munns and Fox, 1976; White and Taylor, 1977; Friesen *et al.*, 1980), or not affect (Reeve and Sumner, 1970; Cabala and Fassbender, 1971; Jones and Fox, 1978) soil P availability as measured by various soil test techniques. Similarly, pH increases have been reported as increasing (Truog, 1953; Jackson *et al.*, 1967; Amarasiri and Olsen, 1973; Ryan and Smillie, 1975; Jones and Fox, 1978; Friesen *et al.*, 1980), decreasing (Juo and Uzu, 1977; Lanyon *et al.*, 1977; McLean and Ssali, 1977; Sumner, 1979a; Farina *et al.*, 1980), or not affecting P recovery by plants (Brams, 1971). Generally, earlier literature on the topic has indicated probable beneficial effects on P solubility and crop recovery at near neutral pH levels (Truog, 1946; 1953), while more recent literature suggests that where P–lime interactions are positive, the effects are more likely at salt pH levels below about pH 5.0.

The most important single publication propounding the benefits of near neutral pH levels is undoubtedly that of Truog (1946). His schematic presentation indicating enhanced P availability near neutrality is still widely accepted in many modern soils texts (Foth and Turk, 1972; Brady, 1974; Donahue *et al.*, 1977; Thompson and Troeh, 1978; Follett, *et al.*, 1981) and certainly continues to influence the thinking of agriculturists. In a subsequent article dealing specifically with the effects of lime on P availability Truog (1953) reinforced his earlier views with experimental data obtained from field studies and concluded by stating that at a neutral pH "the amount needed of locally applied phosphate will be less." The field evidence available to Truog, however, was confounded by the presence of legumes in the rotation and it is conceivable that "typical lime × P interactions," in which P responses in clovers have been observed to disappear after liming soils from pH 5.4 to 6.0 (Jackson *et al.*, 1967), influenced the responses of subsequent crops (maize, oats, barley, and wheat). There is clear evidence that P–lime interactions in many legumes may be related to the effects of pH on nodulation and N nutrition rather than to direct effects on P availability (Helyar and Anderson, 1971). Cooke and Knight (1971) and Jaakkola *et al.* (1977) have more recently indicated possible enhancement of P uptake by small grains at elevated pH levels, but in the former case increased P absorption was associated with marked increases in N uptake and in the latter case, the significance was regarded as uncertain.

Although crop specificity is a complicating factor, the bulk of modern evidence indicates that P–lime interactions in most field crops are intimately associated with active soil Al. It is well known that toxic levels of available Al result in severely restricted root systems and enhanced P recovery is a rather obvious consequence of the improved root proliferation which accompanies liming. It has also been shown by several workers that Al results in the immobilization of P both on and in plant roots (Foy *et al.,* 1978) and this P fixation mechanism undoubtedly plays a role in determining the effects of lime on P nutrition. In addition, although there is disagreement about its significance (Adams, 1980), especially under field conditions, P availability in the soil solution may be reduced by coprecipitation with soluble Al (Syers *et al.,* 1971). At the higher P rates commonly required in pot studies this effect, as reflected by depressed exchangeable Al levels, may be quite marked in certain soils (Shoop *et al.,* 1961; Farina *et al.,* 1981). Phosphorus adsorption on Al and Fe oxides and hydroxides is not considered to be meaningfully affected by liming (Sanchez, 1976), but evidence is accumulating to suggest that the increased adsorptive surface resulting from the precipitation of Al hydroxides following liming may, at least temporarily, increase total P sorptive capacity at salt pH values near 5.0 (Amarasisi and Olsen, 1973; Costigan *et al.*, 1982; Haynes, 1982). This effect would probably be more marked in soils containing large quantities of exchangeable Al relative to the amorphous and crystalline P adsorptive surfaces (Syers *et al.*, 1971) and appears to decrease as pH levels are elevated further. Its significance under field conditions is currently unknown. There are, thus, sound reasons to suggest that significant P–lime interactive effects are likely to be evident at salt pH values below 5.0. The practical significance of these effects individually warrants further examination under field conditions.

While earlier views that P availability is enhanced at near neutral pH levels (Truog, 1946, 1953) now seem questionable, there are several reports implicating reduced P availability in yield depression at such pH levels (Pierre and Browing, 1935; Fox and Plucknett, 1964; Lowther and Adams, 1970; Kamprath, 1971; Lanyon *et al.,* 1977; Sumner, 1979a; Farina *et al.,* 1980a, b, 1982). In greenhouse studies increased P applications have been observed to counter the depressive effects of lime on yield (Sumner, 1979; Farina *et al.,* 1980b) and reduced P absorption has been ascribed to the formation of sparingly soluble calcium phosphate (Kamprath, 1971). More recently, increased Al uptake at near neutral pH levels has been associated with the reduced P content of above-ground plant parts (Farina *et al.*, 1980b, 1982). The mechanisms involved, however, are unclear and it is conceivable that depressed P absorption is the result of complex high-order interactions, as such depressions have also been associated with micronutrient deficiencies (Naftez, 1937; York *et al.*, 1954; Reeve and Sumner, 1970; Kamprath, 1971; McLean, 1971; Spain *et al.*, 1974; Friesen *et al.*, 1980), depressed Mg availability (Sumner *et al.*, 1978; Grove, 1980; Grove *et al.*, 1981) and other factors (John *et al.*, 1972; Christensen *et al.*, 1973; Juo and Uzu, 1977).

Data from both greenhouse and field studies with maize are used here to illustrate some of the P–lime interactive effects discussed above. In Figure 8 interaction between P and lime in the greenhouse study is clearly evident. Both yield and P recovery per pot increased markedly as the salt pH was increased from 4.1 to 5.4, applications of 160 ppm P at the optimum pH resulting in yields superior to those obtained at 320 ppm P on the unlimed soil. The P recovery per pot over all P levels increased from 3.5 mg per pot at pH 4.1 to 7.8 mg per pot at pH 5.4. Similarly, response to lime, almost negligible at low levels of P, increased rapidly as

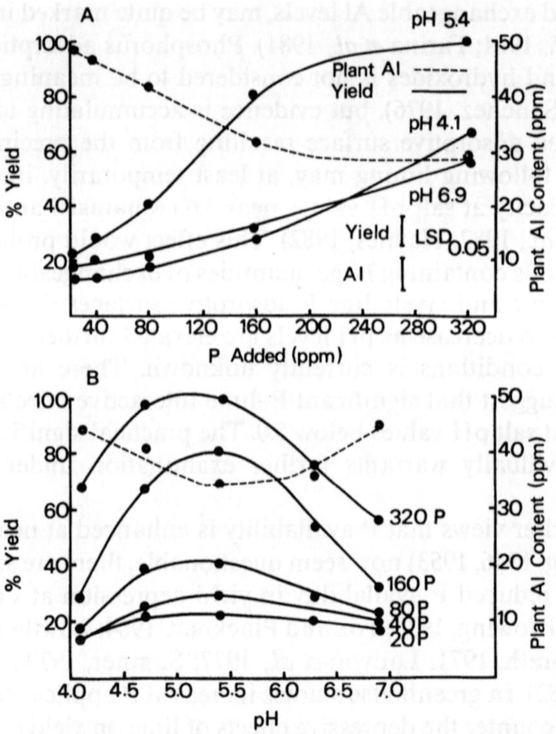

Figure 8. Effect of P added (A) and lime (B) on the relative yield and Al content of maize grown in the greenhouse on a Natal Oxisol. Drawn from unpublished data of Farina.

the rate of P application was raised. The primary mechanisms involved are hard to establish unambiguously but the highest P rate decreased KCl exchangeable Al by 18% in the absence of lime (Farina *et al.*, 1981). Presumably, this reflected a lower P availability. In addition, both lime and P applications resulted in significant reductions in plant Al content (Figure 8), due probably to both soil and plant effects, and root growth was improved markedly by lime and to a lesser degree by P. Thus, several of the possible mechanisms discussed could conceivably have played a role.

The influence of near neutral pH levels was, in this instance, also dramatic. In Figure 8A a close similarity between P response at pH 4.1 and pH 6.9 is evident and the average plant content of Al rather closely reflects the yield response to lime (Figure 8B). Average P uptake per pot dropped from 7.8 mg at pH 5.4 to 3.8 mg at pH 6.9. In terms of content the difference was small (0.01%) but, nevertheless, statistically highly significant and identical to the difference noted between plants at pH 5.4 and pH 4.1. The small differences in plant P content relative to the magnitude of the yield responses to lime and P are difficult to explain satisfactorily, but dilution effects are probably involved. It is also possible, since the P and lime effects on plant Al were additive (independent) and P reduced plant Al content equally at all lime levels, that plant P analysis reflected both metabolically active P and P immobilized by Al. In the case cited here high P applications did not alleviate the depressive effects of lime, but tissue P content did not exceed 0.16% and such an effect would not be expected at markedly deficient P rates. While this evidence in no way precludes the role of other mechanisms in lime-induced yield depressions, it reinforces claims that disturbed P nutrition is sometimes implicated.

Because of the inherent errors associated with field studies and the confounding effects of numerous uncontrolled variables, experimental findings are seldom as consistent and clear cut as those obtained in greenhouse trials. It is, moreover, difficult to duplicate responses equal in magnitude to those commonly obtained in pot experiments and the chances of statistically missing interactive effects are very real. With this in view, it is perhaps reasonable to assume that the field responses of maize to lime and P illustrated in Figure 9 reflect interactive effects akin to those discussed above. Although the interactive effects on yield were statistically nonsignificant, very similar trends are clearly evident. No effect of P on exchangeable Al was measurable, but plant P content in unlimed plots was significantly lower and at the lowest level of P application was actually minimal at the highest level of lime. It is also interesting to note that high P applications tended to reduce the depressive effects of lime so marked at low P rates, an effect compatible with the findings of Sumner (1979a) and Farina *et al.* (1980b) in pot studies.

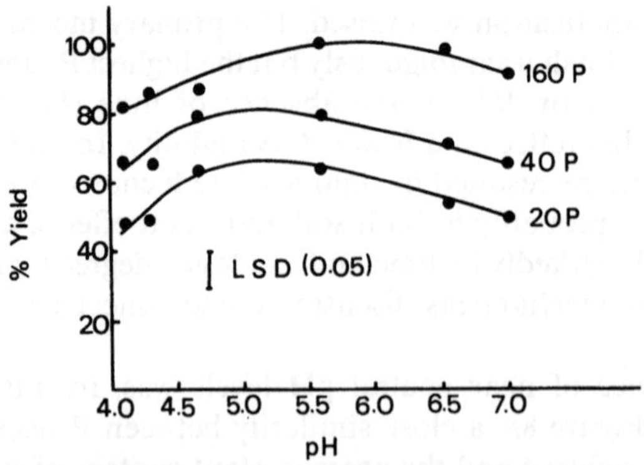

Figure 9. Effect of lime and P on the relative yield of field maize grown on a Natal Ultisol. Drawn from unpublished data of Farina.

While the interpretation of P-lime interactions offered here has undoubtedly not eliminated confusion that surrounds the topic, it has perhaps helped to illustrate why reported results in the literature should appear so contradictory. Generally, there appears to be good agreement that the effects of P and lime may be synergistic in acid soils rich in active Al, because of their common effects on the uptake of Al by plants. However, the involvement of numerous other factors makes consistent experimental demonstration of such interactions difficult. For example, the fact that coprecipitation of P by exchangeable Al is dependent on soil properties other than the pool of exchangeable Al per se (Syers *et al.*, 1971) means that conflicting results in soil solution studies are to be expected. Similarly, the fact that P rates used in pot studies are far higher than those used in field experimentation makes it improbable that pot findings will be easily reproduced under field conditions if, in fact, very high P rates are required before coprecipitation effects become meaningful, as has been indicated in some studies (Farin *et al.* 1981). There are also few field experiments in which the P stresses imposed in pots are approximated and differences in response not unlike the pot and field data illustrated here (Figures 8 and 9) are to be expected. This, perhaps, partly explains the absence of interaction in field maize and peanuts noted by Brams (1971). Yet another complicating factor is the genetic tolerance to high Al and low P noted by Foy *et al.* (1978). Yield depressions at near neutral pH levels are also controlled by imperfectly understood factors, which make prediction of a negative P-lime interaction difficult or impossible at present.

Notwithstanding the conflict that exists regarding P-lime interactive effects at low pH levels, current evidence refuting earlier views that P availability is maximized near neutrality is overwhelming and there can be little doubt that such views require serious reexamination. The economic implications associated with such a revision are extremely meaningful, especially where high-pH-requiring legumes do not form an integral part of the cropping system. It is not uncommon for acid tropical and subtropical soils to require in excess of 30 tons of lime per hectare to bring the pH to the near neutral levels often considered optimal, whereas only a fraction of this amount is usually required to remove active Al.

VI. Phosphorus × Silicon

Silicates react with soil much as lime does and resultant pH changes in highly acid soils can be expected to have similarly beneficial effects on P nutrition. In addition, there is a large body of evidence to indicate that Si per se, independent of any pH effect, results in increased P availability in many soils. These effects and the mechanisms probably reponsible have been comprehensively discussed by Adams (1980), Sanchez and Uehara (1980), Silva (1971), and others and little further can be contributed here in this regard.

Silicate slags produced as a byproduct in steel manufacture are the main economically feasible source of Si. Such products vary considerably from one source to another, may contain significant quantities of potentially harmful elements, and generally have a low reactivity unless finely milled. Low reactivity probably accounts for the fact that silicate use is not widespread, as its competitiveness with conventional liming materials is reduced by further processing costs and higher handling and transport costs often make the use of unprocessed slags economically questionable.

A point of possible significance has been raised by Pearson (1975). There have been persistant reports indicating that the negative P-lime interactions often apparent at elevated pH levels are either absent or greatly reduced when soil pH values are raised to similar levels with silicates (Conner, 1921; Midgley, 1935; Sherman *et al.,* 1964). The reasons are not clear, but it is conceivable that the influence of Si on P and Al availability is involved. Peaslee and Frink (1969) have demonstrated the effect of Si per se on Al availability and it is reasonable to assume that where increased Al uptake has been associated with overliming damage (Farina *et al.,* 1980b), increased concentrations of Si in the soil solution may reduce plant absorption of Al. Further research to establish the reasons for marked differences in crop response to silicates and conventional liming materials in soils limed to neutralility might contribute significantly to our understanding of negative P-lime interactions. It is ironic that an effect so clearly documented over four decades ago (Midgley, 1935) has received so little further attention.

VII. Phosphorus × Iron

An exhaustive review of the literature reveals that there are few if any well-documented cases of interactions between phosphorus and iron in cropping systems in the field. Most of the research work conducted has been confined to solution culture, with a few experiments conducted in soil in the greenhouse. The interactions between phosphorus and iron in the latter experiments indicate that the process is not very simple, being complicated both by the level of acidity or alkalinity in the soil material and by the nature of the rhizosphere surrounding the roots of the particular crop. For any given species, there appear to be cultivars that are efficient in utilizing iron under a set of circumstances where others are inefficient and show iron stress. In general, iron deficiency or stress that may occur in a field crop is much more likely to result from elevated pH rather than from high phosphorus levels. However, both in combination would obviously aggravate the situation.

The only cases in which adequate data have been presented that can be used to verify the model presented earlier are those of Ajakaiye (1979), for millet (*Pennisetum americanum*) and sorghum (*Sorghum bicolor*), and Wallace *et al.* (1976), for beans in nutrient solutions. Their data, replotted in Figure 10, clearly indicate the antagonistic effects between P and Fe under controlled conditions. However, if such experiments were to be repeated in the field, similar results would not necessarily be obtained because of the mitigating effects of other factors that are not readily controlled.

The great affinity between Fe and P both in the soil and in the plant can severely complicate any attempts to explain mechanistically the P–Fe interaction because the insoluble products formed can precipitate or be chemisorbed in the soil after addition or during the translocation and assimilation process.

As far as the importance of P–Fe interactions in field cropping systems is concerned, they will seldom be of great significance as they are only likely to occur at elevated pH levels, which in most agricultural systems should be avoided. Crops grown in naturally alkaline soils often suffer from problems with iron more as a result of the elevated pH than from reactions with phosphorus in the soil.

The literature in this area is totally deficient in reports of factorial experiments in which P and Fe have been studied at a number of levels, which is essential in quantifying interactive effects.

VIII. Phosphorus × Copper

Reports in the literature indicate that this interaction can be antagonistic in some instances, while it can be synergistic in others. For example, in

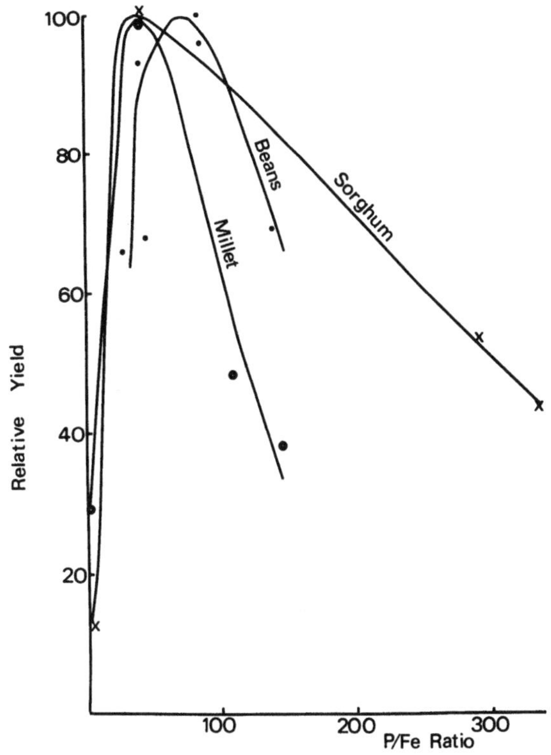

Figure 10. Relationship between yield of greenhouse-grown millet, beans, and sorghum and the P/Fe ratio in tissue. Drawn from data of Ajakaiye (1979) and Wallace *et al.* (1976).

citrus it has been shown by a number of workers (Spencer, 1966; Forsee and Neller, 1944) to be an antagonistic interaction, whereas Greenwood and Halsworth (1960) have shown a synergistic interaction between P and Cu in clovers. However, the literature is almost totally devoid of well-documented references to P–Cu interactions in field cropping systems. The only data that are available come from an experiment conducted with wheat in the field (Touchton *et al.*, 1980). Their data, part of which was supplied on request, has facilitated a test of the applicability of the model to the P–Cu system under field conditions for two consecutive years. In both cases (Figure 11) the relationship between grain yield and the P/Cu ratio of tissue conforms to the model. Similar verification of the model can be obtained from the data of Spencer (1966) for citrus and from those of Shukla and Singh (1979) for wheat grown in the greenhouse. The data of *Tähtinen* (1976) for oats grown to maturity in the greenhouse substantiate the model in terms of the relationship between soil Cu and grain yield (Figure 12). Unpublished observations by the

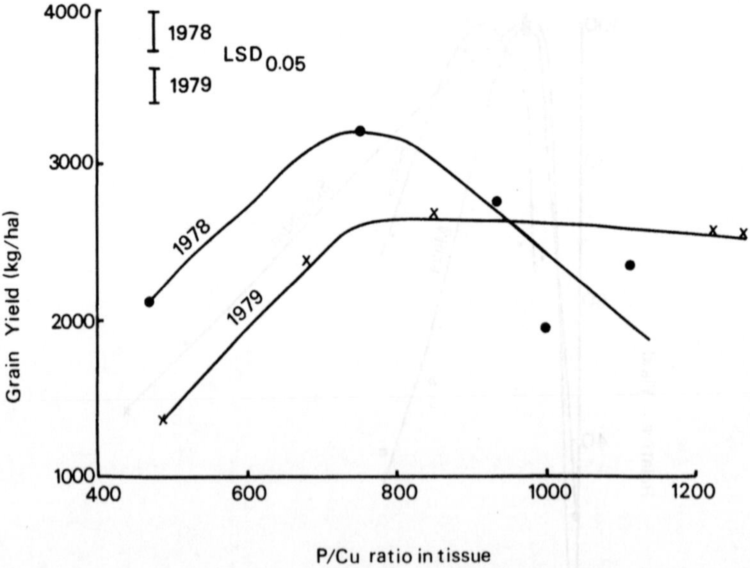

Figure 11. Relationship between field wheat grain yield and the P/Cu ratio in tissue. Drawn from the data of Touchton *et al.* (1980).

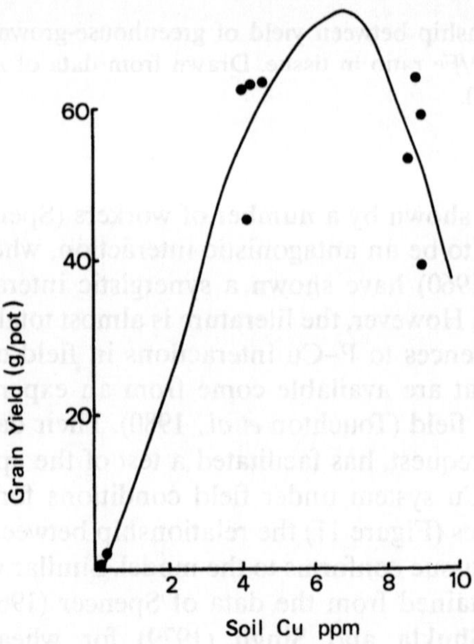

Figure 12. Relationship between oat grain yield (greenhouse) and extractable soil Cu. Drawn from the data of *Tähtinen* (1976).

senior author indicate that Cu insufficiences often show up in corn at production levels in excess of 14 Mg ha^{-1}, particularly on soils having high organic matter contents. Whether or not this type of insufficiency is aggravated by P has not yet been clearly established, because most of the soils being studied are already at high levels of soil P, allowing little chance of studying P–Cu interactions by soil applied treatments.

IX. Phosphorus × Potassium

In his extensive review Adams (1980) concluded that "too little progress has been made in this area (*P–K interactions*) to propose viable interactions at this time." However, very recent work by Reneau *et al.* (1983) has demonstrated that P–K interactions are important in forage sorghum production. Their data have been reworked and plotted in Figure 13. With the exception of a single point the data fit the interpretive

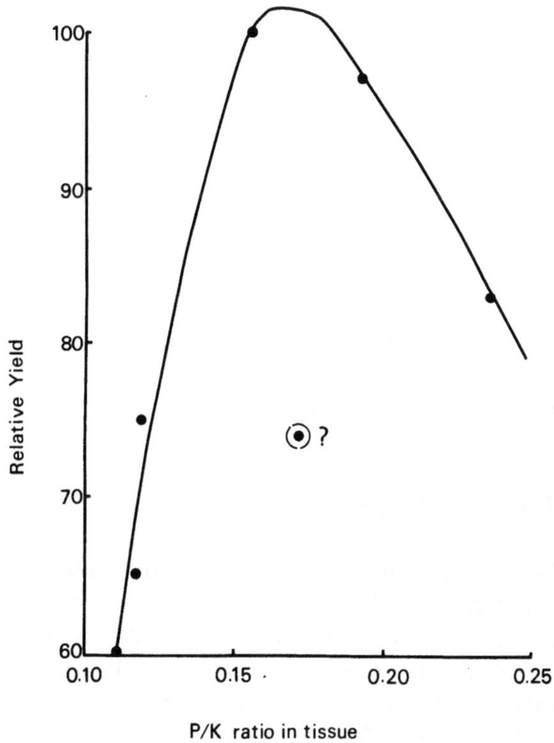

Figure 13. Average relationship between yield of field-grown forage sorghum and P/K ratio of tissue over 3 years. Drawn from the data of Reneau *et al.* (1983).

model well, indicating that the balance between P and K is important. The variant point probably occurs because some factor other than the P/K ratio is determining yield.

X. Phosphorus × Sulfur

Adams (1980) concluded his section on P–S interactions with the statement that "The interaction of absorbed P and S . . . has received scant attention. . . . The lack of attention, however, should not imply a lack of interaction." Such interactions, albeit under greenhouse conditions, have recently been reported by Joshi and Seth (1975) and Kumar and Singh (1980a, b) for wheat and soybeans (*Glycine max*), respectively. The data on the former for the relationship between grain yield and the tissue S/P ratio at 60 days conform to the model. The data of the latter authors were presented in a way that precluded testing of the model. Nevertheless they showed that the balance between P and S was important in terms of producing increased yields, suggesting that their observations would fit the model. Aulakh and Pasricha (1977) investigated P–S interactions in Moong beans (*Phaseolus aureus* L.), which were found to be negative and highly significant in the relation between P and S and the yield and grain quality of the crop.

XI. Phosphorus × Molybdenum

Reports on this interaction under field cropping conditions are not to be found in the literature. However greenhouse and solution culture studies indicate that P generally has a synergistic effect on Mo uptake provided that S is kept at moderate levels, as it has an antagonistic effect (Stout *et al.,* 1951; Singh and Kumar, 1979). However, there is little evidence to show that this synergistic effect on uptake carries through to yield.

XII. Phosphorus × Magnesium

It has been shown by a number of workers (Agbim, 1981; Edwards, 1968; Franklin, 1969) under greenhouse conditions that increased levels of Mg in the root zone favor the uptake of P, which probably is associated with the fact that Mg is an activator for almost all reactions involving phosphate transfer within the plant. Truog *et al.* (1947) showed a positive relation between the P and Mg contents of plants. However, little relevant work has been conducted under field conditions from which quantitative assessments of the importance of this interaction can be made.

XIII. Other Interactions

Little progress has been made in the study of the interactions between P and other nutrient elements since the publication of Adams (1980) treatise to warrant a detailed discussion.

XIV. A Look to the Future

The preceeding discussion clearly indicates that in the case of many interactions of P with other nutrients, little or nothing is known about their importance in the field. The preponderance of laboratory and greenhouse studies in the literature is cause for concern as there is little evidence to show that these types of investigation mirror the behavior of the crop in the field. This trend away from field or "real world" experimentation can readily be understood in the light of the "publish or perish" policy of administrators in research and academic institutions. This is most unfortunate as it is essentially a negation of the mandate of the taxpayers, who require service but whose interests are sacrificed to those of the beauracratic machine. There seems to be a distinct danger of losing sight of the ultimate objective of soil fertility research, namely the investigation of problems facing producers in the field. Unless administrators realize the importance of such objectives, little improvement can be expected in the state of our knowledge.

Notwithstanding the fact that there is a strong lobby and considerable support for field research, too many field investigations are still aimed at the study of single factors at too few levels. There is an urgent need for researchers to adopt the multifactor, multilevel approach in the field by using more complete factorial experimental designs. The study of interactions without factorial designs is impossible. However, the multitude of plots required for factorial designs involving many factors at many levels sets a definite limit on what is possible in the field. New approaches to solving this problem should be sought so that our ability to examine conveniently the effects of several nutrients on yield and one another simultaneously will be improved.

Furthermore, the interpretation of interactions without supporting tissue and soil data is very difficult if not impossible. Even in cases where such data have been collected workers have failed to use the information effectively. This points to the need for a fresh look at diagnostic techniques in soil fertility. An approach recently published (Beaufils, 1973; Beaufils and Sumner, 1976, Sumner, 1977a, b, c, d, 1979b, 1981; Sumner and Beaufils, 1975) goes a considerable way in overcoming some of the drawbacks of the classical approach mentioned above. This technique known as DRIS (diagnosis and recommendation integrated

Table 8. Optimal ratios of P and other nutrients in the leaves of various crops

	Crop		
Function	*Maize*	*Soybeans*	*Sorghum*
N/P	10.68 ± 0.331	—	—
P/N	0.108 ± 0.036	0.069 ± 0.013	0.112 ± 0.021
P/K	0.179 ± 0.083	0.174 ± 0.043	0.259 ± 0.054
P/Ca	0.494 ± 0.231	0.323 ± 0.118	0.795 ± 0.250
P/Mg	1.116 ± 0.641	1.100 ± 0.347	1.518 ± 0.683
P/S	1.362 ± 0.467	1.152 ± 0.305	—
P/Mn	73 ± 73	55 ± 25	40 ± 20
P/Fe	24 ± 9	26 ± 9	20 ± 7
P/Zn	108 ± 61	103 ± 45	180 ± 63
P/Cu	351 ± 226	680 ± 300	560 ± 364
P/Mo	2119 ± 238	124 ± 88	161 ± 86

system) has the ability to handle the interactions of many factors simultaneously. Because of space considerations it is clearly impossible to give an exposé of the system here. However, as an illustration of its potential the norms derived by DRIS for the relationships between P and other nutrients for three crops are presented in Table 8. These norms are in the form of ratios that are means for high-yielding crops together with their standard errors. This range then reflects the variation in compositions that can be considered normal. Values outside the normal range would reflect imbalances resulting from either the numerator or denominator being too high or too low, respectively. Reference to a number of the figures in this chapter will indicate that the norms in Table 8 agree with the values of the particular parameter at which maximum yields in a particular experiment have been obtained. The study and application of this approach to field cropping systems will definitely improve our understanding of the interactions between P and other nutrients.

Frequently seasons interact with the crop's response to soil fertility treatments. Seldom are such interactions ever quantified because of statistical problems. Nevertheless this should not be a deterrent to the promotion of long-term experimentation, which is essential for an improved understanding of interactions.

References

Adams, F. 1980. Interactions of phosphorus with other elements in soil and plants. In: R.C. Dinauer (ed.), *The role of phosphorus in agriculture.* American Society of Agronomy, Madison, WI, pp. 655–680.

Agbim, N.N. 1981. Interactions of phosphorus, magnesium and zinc on the yield and nutrient content of maize. *J. Agr. Sci. Camb.* 96:509–514.

Ajakaiye, C.O. 1979. Effect of phosphorus on growth and iron nutrition of millet and sorghum. *Plant Soil* 51:551–561.

Amarasiri, S.L., and S.R. Olsen. 1973. Liming as related to solubility of phosphorus and plant growth in an acid tropical soil. *Soil Sci. Soc. Am. Proc.* 37:716–721.

Asif, M.I., A.A. Khan, and M.N. Ajakaiye. 1976. Zinc nutrition of onions as influenced by phosphorus. *J. Agr. Sci. Camb.* 87:277–279.

Aulakh, M.S., and N.S. Pasricha. 1977. Interaction effect of sulphur and phosphorus on growth and nutrient content of Moong (*Phaseeolus aureus* L.) *Plant Soil* 47:341–350.

Beaufils, E.R. 1973. Diagnosis and Recommendation Integrated System (DRIS). A general scheme for experimentation and calibration based on principles developed from research in plant nutrition. Soil Sci. Bull. No. 1, Univ. of Natal, Pietermaritzburg, South Africa, 132 pp.

Beaufils, E.R., and M.E. Sumner. 1976. Application of the DRIS approach in calibrating soil and plant parameters for sugarcane. *Proc. S. Afr. Sugar Technol Assoc.* 50:118–124.

Bingham, F.T., and J.P. Martin. 1956. Effects of soil phosphorus on growth and minor element nutrition of citrus. *Soil Sci. Soc. Am. Proc.* 20:382– 385.

Bingham, F.T., J.P. Martin, and J.A. Chastain. 1958. Effects of phosphorus fertilization of California soils on minor element nutrition of citrus. *Soil Sci.* 86:24–31.

Boawn, L.C., F.G. Viets, and C.L. Crawford. 1954. Effect of phosphate fertilizer on zinc nutrition of field beans. *Soil Sci.* 78:1–7.

Brady, N.C. 1974. *The nature and properties of soils.* Macmillan Publishing Company, New York.

Brams, E.A. 1971. Continuous cultivation of West African soils. Organic matter diminution and effects of applied lime and phosphorus. *Plant Soil* 35:401–414.

Cabala, P. and H.W. Fassebender. 1971. Effect of liming on forms and availability of phosphorus in soils of the cacao region of Bahia, Brazil. *Turrialba* 21:36–38.

Christenson, D.R., R.P. White, and E.C. Doll. 1973. Yields and magnesium uptake by plants as affected by soil pH and calcium levels. *Agron. J.* 65:202–206.

Colewell, J.D. 1978. Computations for studies of soil fertility and fertilizer requirements. Comm. Agr. Bur., Farnham Royal, England.

Conner, S.D. 1921. Liming in its relation to injurious inorganic components in the soil. *J. Am. Soc. Agron.* 13:113–124.

Costigan, P.A.,A.D. Bradshaw, and R.P. Gemmel. 1982. The reclamation of acidic colliery spoil III. Problems associated with the use of high rates of limestone. *J. Appl. Ecol.* 19:193–201.

Crooke, W.M., and A.H. Knight. 1971. Crop composition in relation to soil pH and root cation exchange capacity. *J. Sci. Food Agr.* 22:235–241.

Donahue, R.L., R.W. Miller, and J.C. Shickluna. 1977. *Soils—an introduction to soil and plant growth.* Prentice-Hall Inc., Englewood Cliffs, NJ.

Edwards, D.G. 1968. Cation effects on phosphate absorption from solution by *Trifolium subterraneum. Aust. J. Biol. Sci.* 21:1–11.

Ellis, R., J.F. Davis, and D.L. Thurlow. 1964. Zinc availability in calcareous Michigan soils as influenced by phosphorus level and temperature. *Soil Sci. Soc. Am. Proc.* 28:83–86.

Elsokkary, I.H., H.A. El-Attar, and M.A. Amer. 1981. Influence of phosphorus and zinc fertilizer on the uptake of P and Zn by corn plants grown in highly calcareous soils. *Plant Soil* 59:227–236.

Engelstad, O.P., and S.E. Allen. 1971. Effect of form and proximity of added N on crop uptake of P. *Soil Sci.* 112:330–337.

Farina, M.P.W., and P. Channon. 1980. Nitrogen and phosphorus economic optima for maize on a Msinga clay loam. *Crop Prod.* 9:12–16.

Farina, M.P.W, M.E. Sumner, C.O. Plank, and W.S. Letzsch. 1980a. Exchangeable Al and pH as indicators of lime requirement for corn in several widely differing acid soils. *Soil Sci. Soc. Am. J.* 44;1036–1041.

Farina, M.P.W., M.E. Sumner, C.O. Plank, and W.S. Letzch. 1980b. Aluminum toxicity in corn at near neutral soil pH levels. *J. Plant Nutr.* 2:683–697.

Farina, M.P.W., P. Channon, and M.E. Sumner. 1981. A glasshouse comparison of several lime requirement indices for maize. *Crop Prod.* 10:129–135.

Farina, M.P.W., M.E. Sumner, and P. Channon. 1982. Lime induced maize yield depressions in highly weathered soils. *Proc. Ninth Int. Plant Nutr. Coll.* 1:162–167.

Farwell, A.J., P.M.W. Farina, and P. Channon. 1980. The effects of nematodes on nutrient absorption by maize. *Crop Prod.* 9:247–251.

Follett, R.H., L.S. Murphy, and R.L. Donahue. 1981. *Fertilizers and soil amendments.* Prentice-Hall, Inc., Englewood Cliffs, NJ.

Forsee, W.T., and J.R. Neller. 1944. Phosphate responses in a Valencia grove in the eastern Everglades. *Proc. Fla. State Hort. Soc.* 57:110–115.

Foth, H.D., and L.M. Turk. 1972. *Fundamentals of soil science.* John Wiley & Sons, New York.

Fox, R.L., and D.L. Plucknett. 1964. Overliming Hawaiian soils creates problems. *Hawaii Farm Sci.* 13:9–10.

Foy, C.D., R.L. Chaney, and M.C. White. 1978. The physiology of metal toxicity in plants. *Ann. Rev. Plant Physiol.* 29:511–566.

Franklin, R.E. 1969. Effect of adsorbed cations on phosphorus uptake by excised roots. *Plant Physiol.* 44:697–700.

Friesen, D.K., A.S.R. Juo, and M.H. Miller. 1980. Liming and lime phosphorus-zinc interactions in two Nigerian Ultisols I: Interactions in the soil. *Soil Sci. Soc. Am. J.* 44:1221–1226.

Ganiron, R.B., D.C. Adriano, G.M. Paulsen, and L.S. Murphy. 1969. Effect of phosphorus carriers and zinc sources on phosphorus-zinc interaction in corn. *Soil Sci. Soc. Am. Proc.* 33:306–309.

Giordano, P.M., and J.J. Mortvedt. 1978. Response of corn to Zn in ortho and pyrophosphate fertilizer as affected by soil temperature and moisture. *Agron. J.* 70:531–534.

Grove, J.H. 1980. The ion exchange chemistry of soils and liming. Ph.D. Dissertation, University of Georgia, Athens, GA.

Grove, J.H., M.E. Sumner, and J.K. Syers. 1981. Effect of lime on exchangeable magnesium in variable surface charge soils. *Soil Sci. Soc Am. J.* 45:497–500.

Greenwood, E.A.N., and E.G. Hallsworth. 1960. Studies on the nutrition of forage legumes. II. Some interactions of calcium, phosphorus, copper and molybdenum on the growth and chemical composition of *Trifolium subterraneum* L. *Plant Soil* 12:97–127.

Grunes, D.L. 1959. Effect of nitrogen on phosphorus availability In A.G. Norman (ed.), *Advances in agronomy*, Vol. 11. Academic Press, New York, pp. 369–396.

Haynes, R.J. 1982. Effects of liming on phosphate availability in acid soils. A critical review. *Plant Soil* 68:289–308.

Heady, E.O., and J.L. Dilion. 1961. *Agricultural production functions.* Iowa State University Press, Ames, IA.

Helyar, K.R., and A.J. Anderson. 1971. Effects of lime on the growth of five species and aluminum toxicity and on phosphorus availability. *Aust. J. Agr. Res.* 22:707–721.

Jaakkola, A., H. Hakkola, J. Koylijarvi, and P. Simojoki. 1977. Effect of liming on phosphorus fertilizer requirement in cereals and ley. *Ann. Agr. Fenn.* 16:207–219.

Jackson, T.L., E.G. Knox, A.R. Halvorson, and A.S. Baker. 1967. Crop response to lime in the Western United States. In: R. W. Pearson and F. Adams (eds.), *Soil acidity and liming. Agronomy,* Vol. 12, Amer. Soc. Agron. Madison, WI, pp. 26–269.

John, M.K., G.W. Eaton, V.W. Case, and H.H. Chuah. 1972. Liming of alfalfa (*Medicago sativa* L.) II. Effect on mineral composition. *Plant Soil* 37:363–374.

Jones, J.P., and R.L. Fox. 1978. Phosphorus nutrition of plants influenced by manganese and aluminum uptake from an Oxisol. *Soil Sci.* 126:230–236.

Joshi, D.C., and S.P. Seth. 1975. Effect of sulphur and phosphorus application on soil characteristics, nutrient uptake and yield of wheat crop. *J. Ind. Soc. Soil Sci.* 23:217–221.

Juo, A.S.R., and F.O. Uzu. 1977. Liming and nutrient interactions in two Ultisols from Southern Nigeria. *Plant Soil* 47:419–430.

Kamprath, E.J. 1971. Potential detrimental effects from liming highly weathered soils to neutrality. *Soil Crop Sci. Soc. Fla. Proc.* 31:200–203.

Khan, A.A., and G.K. Zende. 1976. Effect of zinc and phosphorus fertilization on the content and uptake of N, P, K, Ca, Mg and Zn by maize and wheat. *Mysore J. Agr. Sci.* 10:574–584.

Kumar, V., and M. Singh. 1980a. Sulfur, phosphorus and molybdenum interactions in relation to growth, uptake and utilization of sulfur in soybean. *Soil Sci.* 129:297–304.

Kumar, V., and M. Singh. 1980b. Interactions of sulfur, phosphorus and molybdenum in relation to uptake and utilization of phosphorus by soybean. *Soil Sci.* 1304:26–31.

Lanyon, L.E., B. Nagshineh-Pour, and E.O. McLean. 1977. Effects of pH level on yields and chemical composition of pearl millet and alfalfa in soil with differing degrees of weathering. *Soil Sci. Soc. Proc.* 41:389–394.

Leece, D.R. 1978. Distribution of physiologically inactive Zn in maize growing on a black earth soil. *Aust. J. Agr. Res.* 29:749–758.

Loneragan, J.F., T.S. Grove, A.D. Robson, and K. Snowball. 1979. Phosphorus toxicity as a factor in zinc-phosphorus interactions in plants. *Soil Sci. Soc. Am. J.* 43:966–972.

Lowther, W.L., and A.F.R. Adams. 1970. The interaction of lime and phosphorus and the nodulation and growth of white clover. *N.Z.J. Agr. Res.* 13:252–262.

McLean, E.O. 1971. Potentially beneficial effects: Chemical and physical. *Soil Crop Sci. Soc. Fla. Proc.* 31:189–199.

McLean, E.O., and H. Ssali. 1977. Effects of phosphorus rate and form in combination with lime and gypsum on yields and composition of German millet and alfalfa from highly weathered soils. *Soil Sci.* 123:155–164.

Midgley, A.R. 1935. Overliming acid soils. *J. Am. Soc. Agron.* 24:822–836.

Munns, D.N., and R.L. Fox. 1976. Depression of legume growth by liming. *Plant Soil* 45:701–705.

Naftez, J.A. 1937. Soil liming investigations. *J. Am. Soc. Agron.* 29:761–771.

Olsen, S.R. 1972. Micronutrient interactions. In J.J. Mortvedt, P.M. Giordano, and W.L. Lindsay (ed.), *Micronutrients in agriculture.* Soil Sci. Soc. Am., Madison, WI, pp. 243–264.

Parfitt, R.L. 1977. Phosphate adsorption by an Oxisol. *Soil Sci. Soc. Am. J.* 41:1064–1067.

Pearson, R.W. 1975. Soil acidity and liming in the humid tropics. Cornell Int. Agr. Bull. 30, Cornell University, Ithaca, NY.

Peaslee, D.E., and C.R. Frink. 1969. Influence of silicic acid on uptake on Mn, Al, Zn and Cu by tomatoes grown on an acid soil. *Soil Sci. Soc. Am. Proc.* 33:569–571.

Pierre, W.H., and G.M. Browning. 1935. The temporary injurious effect of excessive liming of acid soils and its relation to the phosphate nutrition of plants. *J. Am. Soc. Agron.* 27:742–759.

Reeve, N.G., and M.E. Sumner. 1970. Effects of aluminum toxicity and phosphorus fixation on crop growth on Oxisols in Natal. *Soil Sci. Soc. Am. Proc.* 34:263–267.

Reneau, R.B., G.D. Jones, and J.B. Friedericks. 1983. Effect of P and K on yield and chemical composition of forage sorghum. *Agron J.* 75:5–8.

Ryan, J., and G.W. Smillie. 1975. Liming in relation to soil acidity and P fertilizer efficiency. *Comm. Soil Sci. Plant Anal.* 16:409–420.

Safaya, N.M., and B. Singh. 1977. Differential susceptibility of two varieties of cowpea to phosphorus induced zinc deficiency. *Plant Soil* 48:279–290.

Sanchez, P.A. 1976. *Properties and management of soils in the tropics.* John Wiley & Sons, New York.

Sanchez, P.A., and G. Uehara. 1980. Management considerations for acid soils with high phosphorus fixation capacity. In: R.C. Dinauer (ed.), *The role of phosphorus in agriculture.* Am. Soc. Agron., Madison, WI, pp. 471–514.

Sherman, G.D., I.P.S. Dias, and N.H. Monteith. 1964. Calcium silicate, a new liming material. *Hawaii Farm Sci.* 13:8–9.

Shoop, G.T., C.R. Brooks, R.E. Blaser, and G.W. Thomas. 1961. Differential responses of grasses and legumes to liming and phosphorus fertilization. *Agron. J.* 53:111–115.

Shukla, V.C., and Singh, N. 1979. Phosphorus-copper relationship in wheat. *Plant Soil* 53:399–402.

Silva, J.A. 1971. Possible mechanisms for crop response to silicate applications. *Int. Symp. Soil Fert. Eval. Proc.* (New Delhi) 1:805–814.

Singh, M., and V. Kumar. 1979. Sulfur, phosphorus and molybdenum interactions on the concentration and uptake of molybdenum in soybean plants. *Soil Sci.* 127:307–312.

Singh, M., and K.S. Singh. 1979. Response of wheat to zinc fertilization at different levels of phosphorus in loamy sand soil. *J. Indian Soc. Soil Sci.* 27:314–320.

Smyth, T.J., and P.A. Sanchez. 1980. Effects of lime, silicate and phosphorus applications to an Oxisol on phosphorus sorption and iron retention. *Soil Sci. Soc. Am. J.* 44:500–505.

Soltanpour, P.N., F. Adams, and A.C. Bennett. 1974. Soil phosphorus availability as measured by displaced soils solutions, calcium chloride extracts, dilute acid extracts and labile phosphorus. *Soil Sci. Soc. Am. Proc.* 38:22–228.

Spain, J.M., C.A. Francis, R.W. Howeler, and F. Calvo. 1974. Differential species and varietal tolerance to soil acidity. Seminar on Soil Managment Development Process Tropical America Proceedings.

Spencer, W.F. 1966. Effect of copper on yield and uptake of phosphorus and iron by citrus seedlings grown at various phosphorus levels. *Soil Sci.* 102:296–299.

Spratt, E.D., and A.E. Smid. 1978. Yield and elemental composition of flax as affected by P and micronutrients. *Agron. J.* 70:633–638.

Stout, P.R., W.R. Meagher, G.A. Pearson, and C.M. Johnson. 1951. Molybdenum nutrition of crop plants. I. The influence of phosphate and sulfate on the absorption of molybdenum from soils and solution cultures. *Plant Soil* 3:51–87.

Stukenholtz, D.D., R.J. Olsen, G. Gogan, and R.A. Olson. 1966. On the mechanism of phosphorus-zinc interaction in corn nutrition. *Soil Sci. Soc. Am. Proc.* 30:759–763.

Sumner, M.E. 1977a. Preliminary NPK foliar diagnostic norms for wheat. *Comm. Soil Sci. Plant Anal.* 8:149–167.

Sumner, M.E. 1977b. Preliminary N, P and K foliar diagnostic norms for soybeans. *Agron. J.* 69:226–230.

Sumner, M.E. 1977c. Use of the DRIS system in foliar diagnosis of crops at high yield levels. *Comm. Soil Sci. Plant Anal.* 8:251–268.

Sumner, M.E. 1977d. Effect of corn leaf sampled on N, P, K, Ca and Mg content and calculated DRIS indices. *Comm. Soil Sci. Plant Anal.* 8:269–280.

Sumner, M.E. 1979a. Response of alfalfa and sorghum to lime and P on highly weathered soils. *Agron. J.* 71:763–766.

Sumner, M.E. 1979b. Interpretation of foliar analyses for diagnostic purposes. *Agron. J.* 71:343–348.

Sumner, M.E. 1981. Diagnosing the sulfur requirements of corn and wheat using foliar analysis. *Soil Sci. Soc. Am. J.* 45:87–90.

Sumner, M.E., and E.R. Beaufils. 1975. Diagnosis of the NPK requirements of sugarcane irrespective of plant age and season using Beaufils' system (DRIS)—preliminary observations. *Proc. S. Afr. Sugar Technol. Assoc.* 49:137–141.

Sumner, M.E., M.P. W. Farina, and V.J. Hurst. 1978. Magnesium fixation—a possible cause of negative yield responses to lime applications. *Comm. Soil Sci. Plant Anal.* 9:995–1007.

Syers, J.K., T.D. Evans, J.D.H. Williams, and J.T. Murdock. 1971. Phosphate sorption parameters of representative soils from Rio Grande do Sul, Brazil. *Soil Sci.* 112:267–275.

Tähtinen, H. 1976. The effects of lime and phosphorus on copper uptake by oats and on the response to copper fertilization. *Ann. Agr. Fenniae* 15:245–253.

Takkar, P.N., M.S. Mann, R.L. Bansal, N.S. Randhawa, and H. Singh. 1976. Yield and uptake response of corn to zinc as influenced by phosphorus fertilization. *Agron. J.* 68:942–946.

Terman, G.L., and J.C. Noggle. 1973. Nutrient changes in corn as affected by dry matter accumulation with age and response to applied nutrients. *Agron. J.* 65:941–945.

Terman, G.L., J.C. Noggle, and C.M. Hunt. 1977. Growth rate—nutrient concentration relationships during early growth of corn as affected by applied N, P and K. *Soil Sci. Soc. Am. J.* 41:363–368.

Thompson, L.M., and F.R. Troeh. 1978. *Soils and soil fertility.* McGraw-Hill Book Co., New York.

Touchton, J.T., J.W. Johnson, and B.M. Cunfer. 1980. The relationship between phosphorus and copper concentration in wheat. *Comm. Soil Sci. Plant Anal.* 11:1051–1066.

Truog, E. 1946. Soil reaction influence on availability of plant nutrients. *Soil Sci. Soc. Am. Proc.* 11:305–308.

Truog, E. 1953. Liming in relation to availability of native and applied phosphates. In: W.H. Pierre and A.G. Norman (eds.), *Soil and fertilizer phosphorus. Agronomy,* Vol. 4, Academic Press, New York.

Truog, E., R.J. Goates, C.G. Gerloff, and K.C. Berger. 1947. Magnesium—phosphorus relationships in plant nutrition. *Soil Sci.* 63:19–25.

Wagner, R.E. 1979. Interactions of phosphorus in a high yield environment. *Phos. Agr.* 76:45–56.

Wagner, R.E. 1981. Interaction: Name of game is maximum yields. *J. Nat. Fert. Sol. Assoc.* 25:60–74.

Wallace, A., S.M. Soufi, J.W. Cha, and E.M. Romney. 1976. Iron—phosphorus interactions in bush beans. *Comm. Soil Sci. Plant Anal.* 7:101–104.

Warnock, R.E. 1970. Micronutrient uptake and mobility within corn plants in relation to phosphorus induced zinc deficiency. *Soil Sci. Soc. Am. Proc.* 34:765–769.

White, R.E., and A.W. Taylor. 1977. Effect of pH on phosphate adsorption and isotopic exchange in acid soils at low and high additions of soluble phosphate. *J. Soil Sci.* 28:48–61.

York, E.T., R. Bradfield, and M. Peech. 1954. Influence of lime and potassium on yield and cation composition of plants. *Soil Sci.* 77:53–63.

Index

MIX
Papier aus verantwortungsvollen Quellen
Paper from responsible sources
FSC® C105338

If you have any concerns about our products,
you can contact us on
ProductSafety@springernature.com

In case Publisher is established outside the EU,
the EU authorized representative is:
Springer Nature Customer Service Center GmbH
Europaplatz 3, 69115 Heidelberg, Germany

Printed by Libri Plureos GmbH
in Hamburg, Germany